HIGHER EDUCATION IN THE TWENTY-FIRST CENTURY:
ISSUES AND CHALLENGES

PROCEEDINGS OF THE INTERNATIONAL CONFERENCE, AHLIA UNIVERSITY, KINGDOM OF BAHRAIN, 3 – 4 JUNE 2007

Higher Education in the Twenty-First Century: Issues and Challenges

Editors

Abdulla Y. Al-Hawaj
Ahlia University, Kingdom of Bahrain

Wajeeh Elali
Ahlia University, Kingdom of Bahrain

E.H. Twizell
School of Information Systems, Computing and Mathematics
Brunel University, UK

CRC Press is an imprint of the
Taylor & Francis Group, an **informa** business
A BALKEMA BOOK

CRC Press/Balkema is an imprint of the Taylor & Francis Group, an informa business

Typeset by Vikatan Publishing Solutions (P) Ltd., Chennai, India
Printed and bound in Great Britain by Cromwell Press Ltd, Towbridge, Wiltshire.

Published by: CRC Press/Balkema
P.O. Box 447, 2300 AK Leiden, The Netherlands
e-mail: Pub.NL@taylorandfrancis.com
www.crcpress.com – www.taylorandfrancis.co.uk – www.balkema.nl

ISBN: 978-0-415-48000-0 (hbk)
ISBN: 978-0-203-88577-2 (ebook)

Higher Education in the Twenty-First Century: Issues and Challenges – Al-Hawaj, Elali & Twizell (eds)
© 2008 Taylor & Francis Group, London, ISBN 978-0-415-48000-0

Table of Contents

Higher Education in the Twenty-First Century: Issues and Challenges – Al-Hawaj, Elali & Twizell (eds)
© 2008 Taylor & Francis Group, London, ISBN 978-0-415-48000-0

Introduction

There is no doubt that a higher education system is a great asset and brings enormous benefits to both the individual and the nation. The skills, creativity, and research developed through higher education are a major factor in any society's success in creating jobs and advancing prosperity. Universities and colleges play a vital rôle in expanding opportunity and promoting social justice. The benefits of higher education for individuals are far-reaching. On average, graduates get better jobs and earn more than those without a higher education.

In spite of all the progress that has been achieved in the last two decades or so, higher education is still under a great amount of pressure, and at risk of decline. The challenges are growing exponentially and many countries are today facing hard choices on funding, quality, and management. Tackling these challenges needs, among other things, a long-term strategy for investment and reform as well as better infrastructure, better collaboration, and stronger links with business and industry.

IMPORTANCE

In general, this conference will provide important opportunities for academics, policy makers, senior administrators and industrial professionals to interact and share ideas about the existing and emerging practices and strategies that would enhance quality, productivity and innovation, as well as promote social responsibility in higher education.

GOALS

The main objective of the conference is to create a forum in which leading regional and international universities can meet, explore, discuss and develop practical approaches to attain strategic success and co-operation. Moreover, the conference is intended to provide a platform by which all private universities in the GCC countries can join together to form a regional association that will be hosted and based in the Kingdom of Bahrain.

CONFERENCE MAIN THEMES

- Higher Education: Trends and Challenges;
- Building Capacity for Higher Education and Faculty Professional Development;
- International Education and Strategic Partnerships: a Key to Success;
- Quality Assurance and Academic Accreditation;
- Research in Higher Education Institutions;
- Higher Education Labour Markets.

Higher Education in the Twenty-First Century: Issues and Challenges – Al-Hawaj, Elali & Twizell (eds)
© 2008 Taylor & Francis Group, London, ISBN 978-0-415-48000-0

Welcome message

Abdulla Y. Al-Hawaj
President, Ahlia University, Kingdom of Bahrain

On behalf of Ahlia University, Brunel University, and the GCC General Secretariat, I would like to welcome you to the Kingdom of Bahrain, and to our Opening Ceremony of the International Conference on Higher Education in the 21st Century.

It is indeed an honour and a pleasure for Ahlia University to host this important gathering of distinguished scholars, educators, policy makers, higher education administrators and guests for this momentous International Conference on Higher Education in the Gulf region.

Like many of your own countries that have gone before us to pave the way, we, too, in the Kingdom of Bahrain are striving to assert leadership in the area of higher educational excellence in the Gulf region. The patronage extended to this conference by His Majesty, King Hamed bin Issa Al Khalifa, the King of Bahrain, demonstrates the importance that His Majesty attaches to the continued development of the higher education sector in the Kingdom of Bahrain. We, at Ahlia University, are also proud to mention that since its inception in 2001, the University has been taking the lead in promoting excellence and social responsibility in all areas of higher education (HE). Having been the first privately licensed university in the Kingdom, Ahlia University is in the vanguard of private education in the region. Its leadership in the domestic higher educational milieu is manifest in its having the only accredited Ph.D. programme in the Kingdom of Bahrain in collaboration with Brunel University of the United Kingdom.

I believe that all of you who are participating in this prestigious International Conference on Higher Education in the Gulf region, whether you are educators, policy makers, industry professionals or guests, have an important part to play in order to make known the vital role that higher education has in the world today. Each of you has a valid message to send out—or to disseminate. You have all been assembled in this forum to exchange ideas concerning the mobilization of education, which can generate a regional stock of human capital—the true foundation of economic development and social progress in the Twenty-First Century.

Higher Education institutions spur human resource skills-enrichment, which in turn builds the capacity to compete in a globalized world in which the Knowledge Economy reigns supreme. Prosperity or poverty/ascendance or decline—all hinges on the intellectual underpinning of the nation. Natural resource endowments can be depleted, but human resource endowments, if properly cultivated, serve as the root of a sustainable economy. Regionally, HE institutions, more than at any other time in history before the emergence of the Knowledge Economy, bear a heavy burden and so are charged with playing a decisive role in the development of nations in the Twenty-First Century. It is in this context of the vital role that HE institutions play in the destiny of nations that we hold this important Conference on Higher Education in full recognition of how the creative output of faculty, researchers, staff, and students enrich the human intellect without borders.

That being said, the conference theme focusses on universities rising to meet the ever-changing challenges of higher education. Being receptive to growth opportunities, universities build the capacity of its professionals, whether in the Gulf region or beyond, and permit sharing and collaboration with others in this dynamic, all-encompassing field of HE.

This conference aims to create a better understanding of these and other related issues, as well as how to facilitate the means and methods of achieving overall excellence in each area. All of these prime directives can only add to the overall wealth of human resources of a country. These also have a particular significance for all private universities in the GCC region, as well as for policy makers and funding agencies. The conference aim is also to provide a forum whereby regional as well as leading international universities can meet, explore, and develop practical approaches for their strategic collaboration and integration, which would lead to a mutual benefit for all concerned. Moreover, this conference is intended to establish what we tentatively call the "Association of Private Universities in the GCC countries." Accordingly, the Kingdom of Bahrain is proud to play host to such a value-added Association that will, upon its inception, begin to act as a vital component that will support all its members in their quest for excellence.

Allow me to say, finally, that we remain greatly indebted to His Majesty, King Hamed bin Issa Al Khalifa, the King of Bahrain, for his vision, inspiration, and support, without which this conference would not have been staged. Also, I would like to express my deepest gratitude and appreciation to the Government of Bahrain, especially, to His Highness, the Prime Minister Shaikh Khalifa bin Salman Al Khalifa, and the Crown Prince Shaikh Salman bin Hamad Al Khalifa for their commitment to the ideals of Higher Education. Furthermore, I would like to take this opportunity to thank all of our distinguished sponsors; in particular, the Platinum Sponsors: the National Bank of Bahrain, Ithmaar Bank, and MTC Vodaphone for their understanding and generosity. Allow me also to extend appreciation to the organizing committee as well as to the sub-committees for their dedicated hard work that has made this great event possible.

On behalf of myself and the conference organizers, I would also like to express our thanks to our distinguished keynote speakers who have kindly consented to honour us with a timely and informative discourse, which will set the stage for other related contributions and papers. No, I am not forgetting all of the erudite participants, speakers, and honoured guests. Your attendance here today and your contribution touch us greatly and will most decisively make this conference a success. Ahlia University would, therefore, also like to highly commend all of you for your effort, your life's work and your applied knowledge as seekers of excellence in the field of Higher Education, which is why you are in attendance today.

May this conference be a milestone in the development of Higher Education in our region; may your stay with us be a memorable experience, and may your time here be well rewarded.

Thank you and God bless you!

Higher Education in the Twenty-First Century: Issues and Challenges – Al-Hawaj, Elali & Twizell (eds)
© 2008 Taylor & Francis Group, London, ISBN 978-0-415-48000-0

Participating universities and institutions

Bahrain

- Ahlia University
- University of Bahrain
- Royal College of Woman
- Arabian Gulf University
- Arab Open University
- Birla Institute of Technology
- Kingdom University
- Gulf University
- NYIT
- RCSI—Medical University of Bahrain
- Applied Science University
- Delmon University for Sciences & Technology
- Ministry of Education
- Capital Governorate
- Higher Education Council (HEC)
- Economic Development Board (EDB)
- National Bank of Bahrain (NBB)
- Ithmaar Bank
- MTC Vodafone
- Bank of Bahrain and Kuwait
- Bahrain Training Institute

Saudi Arabia

- ARAMCO
- KFUPM-Saudi Aramco
- Dar-Al Hikma College
- Prince Mohammad Bin Fahd University
- Ibn Sina National College for Medical Studies
- Prince Sultan University
- Arab Open University—Saudi Branch

United Arab Emirates

- Abu Dhabi University
- Ajman University for Science & Technology
- Al-Ain University for Science & Technology
- Al-Ghurair University
- Al Hosn University
- Ittihad University
- University of Wollongong in Dubai
- American University of Sharjah

Egypt

- American University in Cairo

Jordan

- Philadelphia University
- Arabic Association of Private Universities

Malaysia

- International Islamic University

Switzerland

- Webster University

Turkey

- Fatih University

United Kingdom

- Brunel University
- University of Paisley
- University of Westminster

United States of America

- University of Wisconsin-River Falls

Kuwait

- Australian College of Kuwait
- Kuwait Maastricht Business School
- Arab Open University, Kuwait Branch
- American University of Kuwait

Oman

- Caledonian College of Engineering
- Sur University College
- Mazoon College
- Dhofar University
- University of Nizwa

Qatar

- College of the North Atlantic
- CHN University
- Texas A & M University

Higher Education in the Twenty-First Century: Issues and Challenges – Al-Hawaj, Elali & Twizell (eds)
© 2008 Taylor & Francis Group, London, ISBN 978-0-415-48000-0

Acknowledgements

SPONSORS AND SUPPORTERS

The organizers of the International Conference on Higher Education would like to extend their gratitude to the following dignitaries and organizations for their generosity and support:

- Ministry of Education
- Capital Governorate
- Economic Development Board
- Bank of Bahrain and Kuwait
- National Bank of Bahrain
- Ithmar Bank
- MTC—Vodafone
- Gulf Air

Higher Education in the Twenty-First Century: Issues and Challenges – Al-Hawaj, Elali & Twizell (eds)
© 2008 Taylor & Francis Group, London, ISBN 978-0-415-48000-0

Organization

SUPREME ORGANIZING COMMITTEE

- **Professor Abdulla Al Hawaj (Chairman)**
- Dr Wajeeh Elali (Secretary-General)
- Dr Amer Al-Roubaie
- Dr Kailash Madan
- Dr Nabeel Moussa
- Dr Yaseen Lasheen
- Dr Hilal Al Shaiji
- Dr Hussain Daif
- Dr Jamal Al Zaer
- Mr Ali Sahwan
- Ms Ahlam Hassan
- Mr Mohammed Rashid
- Mr Art Jones
- Ms Maria Sabri

SCIENTIFIC COMMITTEE

- **Dr Wajeeh Elali (Chairman)**
- Dr Amer Al-Roubaie
- Dr Richard Cummings
- Dr Jamal Al Zaer
- Dr Sayel Al Ramadan

PUBLIC RELATIONS COMMITTEE

Ali Sahwan (Chairman), Mohammed Rashid, Ahlam Hassan, Najma Rafia, A. Karim Sarhan, Shaikha Aliwat, Mohammed Al Qwaiti, Lulwa Budlama, Amal Abdulla, Ohhood Al Daibaib, Maha Al Dosari, Huda Hazeem, Ahmed Al Manaai, Majid Al Modahki, Abdulrahaman Bumoatai, Fatima Al Kaabi, Mariam Al Manaseer.

LOGISTICS COMMITTEE

Mohamed Rashid (Chairman), Dr Richard Cummings, Mrs Najma Rafia, Mustafa Marhama, Sayed Neama, Saju Stephen, Fatima Al Dhaen, Faheema Abdulla, Bedoor Al-Shamlan, Asma Al-Shamlan, Mariam Ahmed Ahli, Sara Al-Arayyed Huda Al-Shamlan, Reem Al-Ammari, Fadhel Abbas Fakher, Deena Ghassan A. Saleh, Narjis Mohamed Al-Mesbah, Fadheela Mohamed Al-Shehabi, Fatima Isa A.Hussain, Susan A.Jalil Al-Aradi, Tasneem Abdulla Al-Haddad, Shaimaa Mahmood Jawaj, Sarah Jawad.

MEDIA COMMITTEE

- **Dr Hilal Al Shaiji (Chairman)**
- Dr Yaseen Lasheen
- Dr Zuhair Dhaif
- Dr Aziza Abdou
- Mr Yousif Al Benkhaleel
- Mr Mohanad Soliaman
- Mr Mohamed Al Saei
- Mr Farah Awadh
- Ms Jameela Al Nashaba
- Mr Ibrheem Taha

REGISTRATION AND SECRETARIAL COMMITTEE

Ahlam Ali Hassan (Chairperson), Maria Akbar Saberi, Fay Fadhel Al Asfoor, Reem Sultan Al-Ammari, Dana Maher Abul, Safa Jamal, Hanan Khabbaz, Nora Al-Shaaban, Ammar Al-Hawaj, Seham Ahmed A. Rasool, Samah Sameer Al-Saati, Sarah Yousif, Fatima S. Hadi, Shereen Khalid Al-Alaiwat, Wael Al-Mutawa.

Keynote papers and abstracts

Higher Education in the Twenty-First Century: Issues and Challenges – Al-Hawaj, Elali & Twizell (eds)
© 2008 Taylor & Francis Group, London, ISBN 978-0-415-48000-0

International education and strategic partnerships—a key to success

Don Betz
Chancellor, University of Wisconsin—River Falls, USA

ABSTRACT: This is the global century. It is possible for an individual to have more contact and interaction with more of the world's population and cultures than at any time in human history. This is a reality that will increasingly contour educational philosophies, goals and processes from today into the future.

This is a time when nation-states share the global arena with other transnational actors, from international organizations, NGOs, consortia of organizations, educational institutions and individuals, all of whom have the capacity to impact aspects of the unfolding global era. Pierre Teihard de Chardin's early 20th century prophecy appears more relevant today than when it was penned almost 100 years ago. "The age of nations is passed. If we are not to perish, we must set aside our ancient prejudices, and build the earth." International relations and its pantheon of actors have fundamentally changed in the subsequent decades.

This is an era when national institutions and education systems can build their capacities to thrive by making multiple, intersecting connections with one another and other international actors. Curricula should strive to be inclusive and, through multiple pedagogical styles, offer insight into and understanding of the multiple forces shaping this global century.

The paper will place our current pedagogical efforts in this global century context and discuss possible partnerships for education beyond national systems and national boundaries. A common denominator for these prospective collaborative efforts should be effectively addressing the contemporary challenges we face that are transnational in scope, do not respect political boundaries nor are contained by distance.

Effective, collaborative, interdisciplinary education is a key to success in the global century.

It is an honour for me to participate in this international conference focussed on the pivotal role higher education will increasingly play in this, the global century. I bring greetings from the University of Wisconsin—River Falls and our faculty, administration and students as well as from the University of Wisconsin system and the state of Wisconsin, USA, to His Majesty, King Hamed bin Issa Al Khalifa, the King of Bahrain; to the people of the Kingdom of Bahrain; to Ahlia University and the sponsors of this conference; and to the delegates here today. We fully recognize the importance of such a gathering which invites us to put aside parochial perspectives and embrace a global view of the challenges and opportunities before us. The French Jesuit philosopher, Pierre Teihard de Chardin, offered a premonition of our dynamic contemporary international situation. At the height of the nationalist era in the early part of the 20th century, he suggested that "The age of nations is passed. If we are not to perish, we must set aside our ancient prejudices and build the earth". Chardin was forecasting a time when the challenges and opportunities confronting mankind would require increasing collaboration in order to be successfully addressed, and that nation-states alone would not be the sole actors determining the course of global affairs.

We are gathered here in Bahrain from various parts of the globe in the opening years of what will be the first true global century; a time of possibility for the most far-reaching, extensive contact among people on the planet in human history; an age when what we know and how we learn will be more important than where we live; an era, unprecedented in human history, when cross-cultural collaboration may be the difference between economic and civil success and failure. Those connections will be vital to global political and human relations where the challenges defy narrow, boilerplate solutions, and problems do not respect national borders. Each day we, our institutions and societies move deeper into the global century. Are we ready? How do we prepare ourselves, our students and our world for success in this time unlike no other?

Since the early 20th century, the world has witnessed dramatic, unprecedented change. Sovereignty is a revered concept and the province of nation-states, but their co-operation with an increasingly broad array of non-governmental organizations is more accurately depicting international relations in our time. The United Nations and others refer to these NGOs as "civil society". Throughout his ten-year tenure as UN Secretary-General, Kofi Annan openly recognized this global involvement and stated that the energies and efforts of nation-states and non-state actors alike are essential to address the "problems without passports", the issues before the international community that respect no political borders and impact all of us. It will be through a level of commitment unequalled in our time that success on so many fronts may be achieved.

Erik Peterson of the *Center for Strategic and International Studies* forecasts that in the next 20 years we will face "Seven Revolutions", challenges and choices to address inter-twining global issues including demography and strategic resource management, principally food, water and energy, information, technology, integration of the global economy, conflict and governance. Peterson confirms that we are indeed in an historic period of "hyper-peril and hyper-opportunity." Without collaboration, success in managing these forces of change is limited at best.

At the UN's Millennium Summit in September 2000, over 170 world leaders placed development at the heart of the global agenda by adopting the Millennium Development Goals (MDGs), which set clear targets for reducing poverty, hunger, disease, illiteracy, environmental degradation, and discrimination against women by 2015.

At the United Nations in 2000, led by Kofi Annan, this international collaborative process articulated eight issues to serve as humanity's agenda for 15 years. They include:

1. Eradicate extreme poverty and hunger,
2. Achieve universal primary education,
3. Promote gender equality and empower women,
4. Reduce child mortality,
5. Improve maternal health,
6. Combat HIV/AIDS, malaria and other diseases,
7. Ensure environmental sustainability,
8. Develop global partnerships for development.

In September 2005, and widely under-reported in the US media, history's largest assemblage of national and global leaders converged on Manhattan's East Side at the United Nations to convene the Global Summit on Poverty. It was there that the international community took stock of its progress and its prognosis for success.

They were there to assess the Millennium Development Goals' progress, and specifically the cascading and protracted effects of grinding, sustained, implacable poverty, hunger, disease and lack of access to education. The unshakable question remained: Can the world community co-operate just enough not to be its own worst enemy in the age when the most deadly and intractable issues respect no national borders nor are deterred by broad oceans? How do we broaden the coalition of collaboration to effectively address global issues? Can we find common ground in the midst of the cyclone of change that seems to dominate our world and our lives?

Quoting from the preamble of the UN Charter, Annan set the tone for the 2005 Summit by declaring the goal to be "so that they may live in larger freedom". MDGs were to be the result of international collaboration with a focus on discernible, measurable progress by 2015.

The Global Summit on Poverty took first stock of the status of efforts and progress made toward each of the goals. While the results were mixed, it is clear that the MDG issues are critical to long term global stability. Each of the MDGs could be advanced by the efforts of higher education institutions collaborating with one another and with a host of other potential partners.

In September 2005 the UN Human Development Report reported that the international system was at a critical crossroads, and that exceptional efforts should be made to address development issues. It found that deeply-rooted human development inequality was at the heart of the problem.

The choices we make, the collaboration we actively pursue, are critical in securing sustained, broad-based success on these and other global challenges. Among the most useful tools we can utilize and develop in such a time of accelerated and systemic change is education. Education, from pre-kindergarten through Ph.D., is one of the effective and efficient ways we can create the future through the people who will lead it and the values that will inspire them.

Given these realities, higher education should and must champion unprecedented levels of collaboration at every opportunity and with multiple, interconnected partners. Guided by the maxim that "none of us is smarter than all of us", and openly embracing the principle of continuous quality improvement, higher education should be modelling the efficacy of a fresh and vibrant style of collaboration and partnerships. It will take an historic, sustained, united effort and a broad variety of actors to meet the challenge of change in the global century.

Innovative, cross-cutting relationships among colleges and universities can be instrumental in providing the full spectrum of information, perspectives, insights and research needed for continued economic growth and enlightened public policy.

Partnerships between institutions fuelled by collaborating faculty from common disciplines have flourished for many years. What is new in the global century is the unprecedented number of these connections enabled by broadband internet access. Inter-disciplinary teams seeking solutions have also proliferated in this information-rich era. We can do more together and faster than ever before. Tom Friedman, in describing our "flattened" world, confirms that everything is in motion and no-one is in charge.

Higher education is now widely characterized as an essential partner in securing long-term economic development.

Globalization is with us and it is a growing force. It is a time of hyper-peril, hyper-opportunity, the emergence of new international actors, and the rise of civil society. In this scenario, leadership is a fundamental, vital imperative.

Globalization can bring peace and prosperity or war and unemployment. Education is central to crafting globalization as a positive force. We must play a leadership role in making this happen. Globalization requires us to look outward and embrace the demands of change. Especially in the United States, we must take this opportunity seriously. This means a concentration on linguistic and cultural knowledge of our world.

The great forces of globalization, unleashing flows of capital, investment, technology, people and ideas are forcing us to change how we see the world. Educated students and citizens with an informed global perspective have never been more important.

As educators what will we face in the next 20 years?

At this historic juncture of the death of distance, the realm of hyper-connectivity and the persistence of cultural, national, and religious stereo-types, stands the critical role of education in the global century.

All this global change and inter-connection, all of this global "flattening" and compressing dramatically and substantively impact what we teach, whom we teach, how we teach, where we teach and why we teach, as well as how we learn.

University partnerships spanning the globe are proliferating in recognition of the mutual advantages of connecting faculty, students and institutional leaders with one another. These partnerships create fresh connections for learning and cross-cultural familiarity.

And increasingly significant partnerships are arising between higher education and PK-12 educational systems. The two are inextricably intertwined and key factors in the national quest for sustained development. Educational systems must think and plan more clearly in a PK-16 framework to maximize coordination and efficient use of resources. They must "begin with the end in mind" and not be deflected by intermediary and parochial benchmarks.

Educational partnerships continue to proliferate as more private enterprises and public agencies comprehend the potential and salience of synergistic relationships among them.

The strategic partnerships should help produce the next generation of high-tech knowledge workers and researchers important to continuing economic growth and development. Equally, there must be a deliberate initiative to create the reservoir of motivated and committed citizens and leaders. This investment in human capital is emerging as a priority in development strategies. Hagel and Brown remind us that "human talent is the only sustainable edge." Strategic partnerships should agree to invest and develop the full range of human potential. We are inundated by corporate advertising exhorting us to connect, collaborate and succeed. Oil giant Chevron widely advertises that they recognize the "world's most powerful energy is human energy. And we will never run out of it." Collaboration at unprecedented levels will be required to thrive in the global century.

English is playing an historic role in providing the *lingua franca* for cross cultural communication in this global century. This proclivity will continue. At the same time and for a multitude of reasons, every effort must be expended to encourage learning other languages as essential to deeper cultural understanding and insight.

In considering strategic partnerships for higher education, the expanding role of English must be judged as a major factor. Societies which have concentrated on growing its capacities in mathematics and science and who have made significant investments in all dimensions of technology and have not developed English speaking, listening and writing skills are discovering they are at a distinct disadvantage. Some national leaders have decided to address this perceived deficiency by mobilizing resources and attention over a protracted period and produce the next generation of English speakers. Through these campaigns, they are confirming the role of English as a necessary and sufficient global communication tool.

In January 2006 the US Presidents' Summit on International Education convened at the US State Department in Washington, D.C., to discuss international education and public diplomacy. Some 120 presidents and chancellors of US universities and colleges, public and private, attended the meeting which was jointly convened by Secretary of State Condoleeza Rice and Secretary of Education Margaret Spellings. The discussions concentrated on possible joint efforts between universities and the federal government to increase the numbers of international students and scholars visiting the USA, to stimulate interest and instruction in a number of critical languages, among them Arabic, Chinese, Farsi and Korean, and to send more of our own students abroad. All of this activity is seen to be vital to America's well-being. Secretary of State Rice noted in her remarks that many of her colleagues among world leaders were educated in the USA. The result of the Summit was the initiation of the National Critical Languages Initiative which encourages institutions and students to concentrate on increasing the number of students studying these languages.

The assessment of higher education's efficacy is being reframed by technology. "Seat" time replaced by outcomes; teaching by learning.

Partnerships in education during this global century are also aggregating and connecting institutions with national governments, international organizations, inter-governmental and non-governmental organizations. The number and quality of institutional exchanges of faculty and students as well as collaborative research will escalate. Science, mathematics, technology and English will be among the essential variables for success.

For example, South Korea is currently undertaking a realigning course of action to significantly improve English language proficiency throughout the country. The Minister of Education and Human Resources Development recently announced a sweeping plan. The first focus is on English teachers who will have to conduct English classes without help from native English speaking teachers by 2015 according to a recent report. The national government acknowledges that Koreans in general lack strong English skills. To achieve this defining goal, South Korea will be partnering with English-speaking institutions and teachers.

This ambitious plan will require assistance from English-speaking educational institutions, many from other countries. The plans call for the gradual establishment of English immersion programmes from elementary through secondary education at selected national sites. This vision could have a profound impact on South Korea's future as it plans to move the GNP and overall standard of living further up the comparative international scale from its current 12th position. The role of higher education in this dimension of South Korea's economic development confirms the importance of partnerships and the commitment to being competitive among the rapidly changing realities of contemporary economic and political relations.

International organizations such as the World Health Organization (WHO) and the World Bank (IBRD) recognize the efficacy of education in achieving their aims. Their "Building Strategic Partnerships in Education and Health in Africa" initiative has been designed since 2002 to link mainstream education and health priorities for each government.

The partnerships often involve more than two parties. For example, the UK—Indian Education Research Initiative (UKTERI) champions long term links between youth in both countries by providing opportunities for hundreds of young people to start a life-long interaction with each other's country. This initiative is assisted by corporate partners BP, Glaxo Smith Klein and Shell as well as the Association of Indian Universities and the University Grants Commission (UK).

From broad-based efforts such as "Friendship through Education" to the work of organizations such as Sister Cities International and Rotary International, dozens of co-operative agreements and programmes exist and continue to proliferate. Most, if not all, have at least indirect connection with higher education in their own nation-states or with others elsewhere.

Essential priority partnerships for western and especially American higher education institutions should be created and strengthened with institutions in Muslim countries.

It may be important to offer perspectives on teaching and learning about Islam, the Muslim world and the Middle East and to offer a sense of what may seem to inhibit effective cross-cultural connections, and what factors may encourage it. This is an opportunity to identify and articulate a range of challenges and opportunities with which we now live in this globalized world. Education assumes a disproportionately significant role at this juncture.

We in the United States must prepare our students, institutions and our society for constructive relationships with the Arab and Muslim worlds. For that responsibility falls to us more than any other actors in the American national system today. The path cannot be one of "shock and awe". We are struggling against unhelpful stereotypes reinforced by persistent extremism. Such challenges simultaneously elicit the best and worst from everyone.

Forty-three years ago, my first encounter with the Middle East, the Arab world, the Muslim World and Islam was the harbinger of my involvement and connection for the rest of my life. And it happened through higher education.

To contend with some current perceptions are the following recommendations:

1. Take enhanced measures to make our institutions welcoming settings for international students and scholars.
2. Participate in public diplomacy *via* people-to-people contact and interaction with NGOs from Muslim and Arab countries. Intentionally build connections among civil society. These initiatives are more possible now than any other time in history.
3. Study history, especially the rise of Islam and relations with the West from both perspectives.
4. Address stereotypes in re-framing our individual and international relations with the Muslim world.
5. Go to the region; send and take students, faculty and community leaders.
6. Bring Arab/Muslim university presidents, scholars and students to our campuses as well as sponsor speakers, forums and cultural programmes in both directions.
7. Teach Arabic, Farsi and other languages of the Arab and Muslim worlds at our institutions.
8. Visit Arab/Muslim universities. Establish vibrant, productive relationships among presidents of institutions and others.

Today every country is scrambling to find its place in the "flattened world" popularly described by Thomas Friedman. "Every country thinks that it is behind", writes Friedman. "Innovation is often a synthesis of art and science and the best innovators often combine the two." Every country is scurrying to upgrade its human talent base. It is science and mathematics interwoven with art, literature, music and the humanities, the integrated curriculum offered with expertise, professionalism and passion at UW-River Falls and our sister institutions, that best prepare students for international challenges yet unimagined. Investing in our people and their ability to create and to thrive in an integrated global environment will not only lead to productive and fulfilling lives, but also infuse our societies with citizens and leaders who understand opportunity and responsibility.

Our frame of reference will be global. From UW-River Falls' continuing co-sponsorship of the "Wisconsin in Scotland" programme and the "International Traveling Classroom" and "Semester Abroad Europe" programmes to the numerous study tours offered by faculty, UW-River Falls is assisting students in the essential process of developing an informed perspective, a global world view. Exciting possibilities with existing and new international institutional partners virtually on every continent will ensure that our students and faculty will continue to re-discover our world, and bring their new-found insight home, just as international students and faculty discover River Falls to be their home away from home.

Beyond all of its stated goals of teaching, research and service, and its encouraged participation in, and active contribution to, economic development opportunities, higher education seeks partners and is sought by them to shape our societies, both large and small, local and global. Tom Friedman of the *New York Times* wrote last month that this is "…an age when everyone increasingly has the same innovation tools and the key differentiator is human talent". Higher education is dedicated to be the premier developer of that talent.

We also believe that we are here to create and sustain an environment of mutual respect, professional behaviour and an appreciation of individual differences and a celebration of the rich mosaic of human diversity in all its styles, shapes, contours and hues. If this outcome is shared in common, then higher education can be among the most influential of all partners for virtually any enterprise, NGO or government, local or international, as we determine our options by what we do or fail to do in this most inter-dependent of human eras.

Ours is the age of connection and choices, and higher education is emerging as a substantive, universal partner.

Higher Education in the Twenty-First Century: Issues and Challenges – Al-Hawaj, Elali & Twizell (eds)
© 2008 Taylor & Francis Group, London, ISBN 978-0-415-48000-0

Building a university research culture

Chris Jenks
Vice-Chancellor, Brunel University, UK

In an age which is increasingly coming to define itself as organized around 'knowledge economies', we can regard higher education as a growth industry and to a large extent this is a view both driven by and, to a degree, supported by government policy across the developed world. Certainly within the UK higher education is now a costly commodity that our students and their families invest in because of the anticipated returns in terms of employability and life-chances. All UK universities share a three-fold mission to develop research, to educate the population into a broader and more flexible skill set, and to alter the social structure by maximizing the use of available talent and broadening participation. On this last point it is significant that during the 1960s in the UK 5% to 7% of the adolescent population attended university; now it is nearer 40% with a government aspiration to reach 50% participation by 2012. Not all UK universities exercise this tripartite mission to the same degree, however. By far the strongest indicator of a university's status is its degree and quality of research-intensity. All of the highest status ('the best') universities as designated by both national and international league-tables are those who are well known for the quality and sustainability of their research cultures. The quality and sustainability of the research culture in turn creates a positive cycle such that the most highly qualified academic staff and school students apply to the most well known research-intensive universities in the largest numbers, thus the excellence reproduces itself and a quality brand is both established and maintained. All this despite the tendency to regard students as consumers who would, according to market forces, pursue the best teaching—the best delivery of their purchased service. The quality brand, determined by research quality and reputation, always triumphs and research quality and teaching quality are conflated. Nobody ever questions the quality of the teaching at Oxford or Cambridge, for example, even though we might anticipate that it is as humanly variable there as in any other higher education institution.

Now, a good reason for this hierarchy emerging and lasting might be that actually universities are, primarily, institutions that engage in research with teaching being an important but complimentary part of their core business, but not their primary *raison d'etre*. This is a philosophical, political and economic debate that we might conduct elsewhere; the fact remains that universities are primarily differentiated by their research cultures and it is this aspect of their being that I seek to address here.

I think it is important from the outset to stress that building a university research culture is not a once-and-for-all activity; it is a process that requires a concerted strategy, resources and primarily time. University research is always about generating new knowledge and, critically, such knowledge has an inseparable two-fold purpose being: that it should contribute to the economy and that it should contribute to the quality of life. A research culture operates at a number of levels: it is a mind-set of innovation and creativity, an aesthetic of discovery, a perpetual motivation to transcend the present and the conventional; it is also a critical mass of symbiotic thinkers, an appropriate and developing infrastructure, a sustainable income stream and a record of achievement and recognition. All of this points to the notion of a 'tradition', a history of ideas emergent from a history of creative relationships. Tradition can be orchestrated and augmented by university management devising careful strategies that direct behaviour, manage resources, and govern the politics that relate the previous two elements. As previously stated, however, traditions are predicated upon time and development.

The Higher Education system in the UK has a considerable history. Our most established universities like Oxford, Cambridge and Durham are several hundred years old, London University is over 100 years old, the large 'civics', like Leeds, Manchester and Sheffield, for example, were formed in the first part of the 20th century, and the 'modern' universities like York and Kent, Loughborough and Brunel were all established between 40 and 50 years ago—all of these institutions are research-intensive, they all have established research cultures. A large group of 'new' universities, introduced since 1992 with a view to rapidly expanding the national provision of undergraduate education, have pockets of research activity but no recognizable research culture, indeed, many of these institutions do not seek to emulate the 'old' university model. Nevertheless, overall the time dimension is instructive.

Taken as a block the UK higher education system is a success story. The system manifests a high level of research output, an established and growing network of knowledge transfer and enterprise through relations with commerce and industry, a good record of increasing and widening participation across the strata of society, and a high and measurable quality in learning and teaching. Three UK universities rate among the world top ten and many other figure in the top 200 world-renown list.

What these positions reveal and rest upon is our collective ambition to continue to have a higher education system that matches the best in the world. Our HE system has also demonstrated its resilience and its sustained productivity despite periods of significant government under-investment. The system has absorbed a great deal of growth and increased productivity often with a shrinking unit of resource.

Indices of success of the sector, often quoted, are as follows: according to the National Student Survey of 2006 more than 80% of our students are satisfied with their overall experience and their courses; our student non-completion rates are both low and also among the best in the world (in 2003–2004 they were recorded at 14.9%; although the UK has only 1% of the world's population, it carries out 5% of the world's research and produces 12% of all cited papers (the UK ranks second in the world to the USA on research). The United Kingdom's higher education institutions are worth GBP 45 billion (BHD 33.6 billion) to the country's economy on a public investment of GBP 15 billion (BHD 11.2 billion)—a three-fold return.

Let us now look at the components of a research culture and see how this might inform our strategies for developing such a culture to a high quality. Although there will be overlap and convergence I would suggest that there are five key elements to any research culture: first, the academic staff—you are only ever as good as your faculty; second, outputs—the tangible outcomes of the research process and their measurable impact; third, the university environment—which includes funding, students, infrastructures and facilities, and academic activities; fourth, the esteem indicators—the public markers of your success and public evaluation; and finally the ethos of the institution—what I earlier referred to as a mind-set.

We will examine these five elements in sequence but underlying them the development of a successful research culture requires an appropriate context or set of conditions within any university. Primarily a university must provide its members with a stable system, a sustained concentration and focus on core business namely research and knowledge transfer and teaching and learning. Universities can, and of course do, engage in activities outside of core business; they sometimes promote grand schemes and seek publicity through peripheral but high profile developments. All such action, however, must always be secondary to core business. Universities neglect what universities are for at their own peril. They should not compete with different forms of institutions led by different missions, they should not neglect their central purpose as proscribed by funding mechanisms and they must not neglect their contribution to their culture. Though operating in a business-like manner universities are not simply businesses. Following from this primary commitment all universities, like all big organizations, must be directed by strategic planning and such strategic planning must enshrine a moral commitment to the elements of core business, if the development of a research culture is a central goal then this must pre-figure in the planning. The strategy must also have a time-scale and measurable indicators of achievement. The strategy can evolve as the institution develops towards its goals. A further operational necessity is that the university should have a budgetary model that enables rather than inhibits its research drive—can the model enable risk; can the model support interdisciplinary work; can the model pre-invest in research fundamentals? Last, but certainly not least, the development of a research culture requires collective ownership within the university—it is not viable solely as the desire of the Principal or the senior management team; all colleagues, administrative and academic, must share and own the vision for it to become part of their motivation and ultimately a reality. Resistance, disassembly from or apathy towards this shared ownership must be overcome initially through transparent and clear communication.

Let us return now to the five key elements of a research culture previously cited, and to begin with, academic staff. Clearly one critical point for the quality assessment and evaluation of colleagues is at appointment. If the existing reputation of a university is high it will, within that self-fulfilling positive cycle, attract a high quality of applicants and *vice versa*. Wherever a university is located in terms of reputation it is necessary to recruit as far as possible in a selective and highly focussed manner. All new colleagues need to be at least research-active and demonstrably so; that is, they must have a publication record. All new staff should be put on probationary terms that include clauses addressing research productivity; it is important to instil this as a minimum expectation from the beginning of a colleague's career within the institution. Colleagues should become accustomed to publishing at a quality level but also regularly, the ability to deliver on time and the ability to publish regularly are important signs of a future researcher. Having recruited academic staff, given them an induction programme into the strategy of the university, set them research-oriented probationary targets and perhaps paired them with an established staff member as a mentor for their progress, it is essential to provide them with forms of staff development and this should sustain throughout their career. Staff development can have relevance to any aspect of an academic's expected performance, such as learning and teaching, administration or leadership but in trying to build a research culture we must ensure that all colleagues, both new and established, are subject to exercises and training that will improve their skills as a researcher. This might be about prioritizing and time-management, the use of IT, methodological techniques, writing a journal article, finding a publisher, communication and self-presentation or applying for a research grant. Such development must be accessible, improving, inclusive and relevant. No university has unlimited HR resources to throw at staff development but much research training and the sharing of good practice in this regard can be provided in-house at a department of institutional level. Above all it is important to engender a culture where research is the pre-eminent and driving concern of all colleagues and thus an atmosphere that is conducive to such activities. Staff promotion should also have clear and explicit criteria focussed on research productivity. All professorial appointments must gain their credibility from their research output. Finally, staff retention is a serious issue and one that stems from appropriate recruitment, no dislocation between individual and institutional expectations, support and development, suitable research infrastructure and an atmosphere and collective image with which people want to identify. It is better to spend money on staff while they are working with the institution than to offer them a pay rise if they have decided to leave the institution. It is interesting that, if your university has developed a good research reputation, then your staff will be more desirable and competitors will try to poach them. In employment terms there is nothing wrong in developing staff so well that they can launch themselves into the next stage of their career elsewhere. That way you have a healthy turn-over of staff and an opportunity to alter the demography of your faculty.

Secondly, we must look to the research output produced by staff. We have looked above at expectations of productivity levels and that is a significant issue; however, colleagues need to be advised and reminded also about quality issues. Young colleagues do not necessarily know the highest-status journals to aim for, established colleagues may have become accustomed to publishing in 'safer' journals. Quality standards need to be a living presence and as they vary, to some degree, between disciplines, management needs to elect the key performance indicators. Is it impact factors, is it citation rate, is it the status of your publishing house, is it journal *versus* book, single-authored *versus* joint-authored, conference papers, chapters in books, edited *versus* authored and so on? Colleagues need to have this clear or they may waste their research effort through inappropriate publication. In the UK the government, through the funding council (Higher Education Funding Council) conducts a national audit of productivity and quality on a (now) seven-year cycle. This is the Research Assessment Exercise (RAE—the most recent being the RAE 2007) and the results of such audits provide a national and international 'branding' of quality through a numerically-based methodology and subsequently guides the distribution of government Quality Research (QR) funding which is substantial and ranges from as little as 0.5% of some universities' turnover to as much as 30%. The 'branding' provides a metric by which students and staff choose which university to apply for or adhere to, and the funding sustains the level of research activity achieved or not.

Third, we must address what has come to be called the university 'environment'. This is essentially a further set of indicators, both quantitative and qualitative, that demonstrate the degree of research life in the machine, but also the sustainability and longevity of the research culture. Here we have external research funding which may (in the UK) derive from Research Funding Councils that are organized around clusters of disciplines (e.g., the Economic and Social Research Council [ESRC]) or from charities, commercial and industrial organizations. Again, if successful, funding here can be substantial. In my own institution in the UK, Brunel University, external research funding has continued to increase by 11% year-on-year and here we are considering sums of up to GBP 20 million (BHD 14.9 million) per annum by 2012. Achievement in this area also provides the institution with a 'track record' of delivering and reliability and a culture in which staff feel empowered to apply for such funding. The overall pragmatic outcome of such activity is that it sustains and grows the established research culture further. Another major indicator is the existence of post-graduate research (PGR) students and their percentage of the overall student population. In the UK, PGR students are an unregulated source of income for universities but more significantly they demonstrate the maturity and desirability of a university's research culture—the sole criterion for postgraduate choice of institution is the quality and reputation of your faculty. More than this PGR students support the research and publication of the faculty and they generate a tradition of ideas that stems from your institution (it may even develop into a 'school' within a particular discipline). PGR students are, in turn, a tremendous source of advertisement for the quality of your developing research culture. Athird indicator of environment is the sheer scale and dynamism of academic and research activities within an institution as shown by organizing and hosting conferences, running learned symposia within disciplines, and a regular pattern of seminars, workshops and visiting speakers. Such activities demonstrate a sustained level of academic interest and the institution's ability to attract interest and significant individuals from the outside. Needless to say a university environment is also supported and improved by material factors such as the quality of the research infra-structure and the learning and teaching facilities. However, and I would not want to push this point too far, faculty members and PGR students adhere more to intangible qualities of a research culture and less to state-of-the-art facilities.

A further significant dimension comprising a university research culture stems from 'esteem' indicators. These will vary according to the broader context within which any university finds itself according to national conventions and practices but I shall cite those from the UK and anticipate a high degree of similarity. Again we need to stress in this area that academic colleagues need to be encouraged to actively participate as such participation often carries an extra burden. Are your faculty members of national bodies, do they contribute to policy making within education and beyond—are their skills and expertise part of public debate? How far are your faculty members engaged in the organization of academic publishing—do they act as readers for publishing houses; are they editing series for publishing houses; how many actually edit journals or sit on the editorial board of journals—are they, to this degree, gatekeepers for their profession? How many of your colleagues, both nationally and better still internationally, have been awarded prizes for their work, honorary fellowships, honorary professorships or even honorary degrees? To what degree do members of your faculty contribute to funding bodies or work alongside funding organizations in an advisory capacity? Finally to what degree do all colleagues value and pursue making contributions (and particularly keynote speeches) at national and international disciplinary and inter-disciplinary conferences? Above all, we managers need to note that the achievement of esteem indicators can be both tutored and incentivized.

All of the above factors are elements within a holistic view of what serves to develop and what provides a totality around the notion of a research culture. Ultimately, and as a collective form they both generate and cohere around a university ethos, a moral bond, which recognizes the priority of research, which facilitates the integration of research and teaching, and which manifests a sustained commitment to the generation of new knowledge.

In conclusion we can all note that higher education cannot remain trapped within national boundaries. The future of higher education lies in global education and the future of knowledge economies lies in globalized knowledge economies. Strong and vibrant universities seek to form structured and sustainable partnerships abroad in research; in enterprise and consultancy; in knowledge transfer; in the provision of overseas programmes; to establish feeder institutions; and to facilitate staff and student exchanges.

At my university, Brunel University, UK, these imperatives are enshrined in our new strategic plan. We are a creative community that seeks to shape the future;

- We are innovative,
- We are entrepreneurial,
- We are research-intensive,
- And we are output driven.

Higher Education in the Twenty-First Century: Issues and Challenges – Al-Hawaj, Elali & Twizell (eds)
© 2008 Taylor & Francis Group, London, ISBN 978-0-415-48000-0

Challenges facing the privatization of higher education in the Arab World

Amin A. Mahmoud
Former Minister of Culture, Hashemite Kingdom of Jordan and Secretary General, The Association of Arab Private Institutions of Higher Education

ABSTRACT: After defining the concept of privatization of higher education, this paper presents trends in private educational institutions in the Arab World with a view to pinpointing major challenges besetting privatization. These trends include: blurring of the line between public and private education, development of innovation in education through privatization, competition among private educational institutions to improve quality continuously, and the fluidity of development of private educational institutions in different Arab countries. Throughout the Arab World, private education is a necessary consequence of the inability of public education to meet the volume of demand for higher education in the Arab World and a realization that monopolization of higher education by the state runs contrary to national interest. The author concludes with an advocacy of co-operation, even partnership, between private and public universities in the Arab World.

1 INTRODUCTION

This paper attempts of define the concept of "privatization" in higher education; it reviews the various trends that are taking shape in the so-called private higher education institutions in the Arab world, and pinpoints some of the major challenges, which are being felt already by these institutions.

It is a fact that most Arab public universities today are unable to meet the demand for higher education, both in terms of numbers and in terms of quality. They are heavily subsidized and run at considerable financial loss. Their avenues (from government and limited student fees) remain roughly the same as before and cannot satisfy their desire for expansion and development. It is also a fact that most of these universities are intensely overcrowded and, consequently, cannot possibly absorb all students who wish to enroll in them (estimated at some 6.2 million students by 2010).

Therefore, the emergence of private universities was an inevitable development, and a necessary one, to assist in solving the problem of increasing demand for student places. In some Arab countries, private universities and higher education institutions managed at one time to take over 40% of total student enrollment (Jordan). In some Far Eastern countries, such as Japan and South Korea, enrollment percentage of private higher education exceeds 50%, while in most Western European countries private higher education is still strongly resisted. In the USA, private higher education is around 20% of the total enrollment, but is increasing fast.

Fourteen Arab countries have officially licensed private universities. These are: Iraq, Jordan, Lebanon, Egypt, UAE, Yemen, Sudan, Palestine and Morocco; Syria, Saudi Arabia, Oman, Kuwait and Bahrain are following suit.

At present, there are more than 150 private universities and university colleges in the Arab World. This represents some 41% of the total number of Arab universities.

Consequently, private university education has already taken a firm hold and is playing an increasingly active role in the development of higher education in Arab World. So far, however, most of these private universities are run on a purely commercial basis. They do not receive government funds and rely almost totally on student fees to cover all their expenses and leave a margin of profit, if any, for their shareholders. The motivation for profit is thus competing with that of quality. Under these conditions, most, or all, are not interested in supporting research or graduate studies, as their main concern is to increase the number of paying students and minimize costs. Therefore, these universities require serious improvement in their standards and should be subjected to more rigorous accreditation requirements and controls. There are many people who are particularly against certain types of privatization in higher education institutions, especially those owned by one owner, or by those who are clearly and solely profit motivated.

The biggest source of confusion and obstacle to progress is the lack of common quality standards by which the performance of all academic programmes, private and public, can be measured. As in every important industry, establishing common standards of performance and quality assurance under the supervision of independent bodies or agencies, is an essential pre-requisite for creating a dynamic environment of healthy competition among all education providers. Consequently, after decades of quantitative expansion (without a significant increase in the role assumed by the private sector in the provision of HE), the establishment of an independent system for comparative evaluation of the quality of academic programmes in the various fields appears to represent the most natural and high priority goal for the next phase of development. This will bring consideration of quality (and not just quantity) into funding decisions by public and private authorities. It will also lead to improved compatibility between funding decisions on the one hand and national and market priorities on the other hand.

This step will inform decisions by students, employers and donors with regard to the comparative quality of various programmes in each field. It will thus create a healthy competition between academic programmes in all universities and in every field. This will replace confusion and negative competition that are created by conflicting self-promotional claims and counter-claims that are not supported by real evidence by neutral assessments tests.

New sources of HE funding (through grants and endowments) from government, industry, national, international, and philanthropic donors, for programmes of proven high quality in public or private universities would be generated. It is much better to attract additional funding for high quality programmes than through creation of parallel programmes of dubious purpose and quality.

Governments should take measures to reduce their control and interference in the affairs of universities, allowing them more freedom to compete and enhance their own revenues. The measures should be based more on quality criteria, which should guide governments' funding decisions. The role of government will thus be generally transformed from that of the controller and funding provider to that of a regulator and selective financial supporter. Quality assurance could thus become the cornerstone in the process of HE reform.

2 THE CONCEPT OF PRIVATIZATION IN HIGHER EDUCATION

As a phenomenon and a system of thinking, privatization in higher education is not a new development in the Arab World. In fact, some universities are relatively old. Al-Azhar University, for example, was, when established, totally dependent on private donations, grants and Islamic endowments.

The Egyptian University of Cairo was established in 1908 by private groups before it was 'nationalized' in 1925. Similarly, in Lebanon, several higher education institutions, albeit run by foreign establishments, were established around the middle of the 19th century, including the American University of Beirut and several missionary colleges.

However, with the advent of independence movements in the Arab World since the 1950's, most Arab states set up their own government-run universities. Indeed, it is interesting to note that nearly 75% of all Arab Universities were established during the last quarter of the 20th century.

Today, there are some 300 universities and more than 600 colleges with a total enrolment of some 3.5 million students (compared with 15 million students in the U.S.A in 2005).

The number of Arab private universities has grown steadily from a scant few in the 1980's to more than 120 in 2007. However, the total number of university students is still relatively low; it is now around 12.5% of the population compared with around 60% for some developed countries.

Although the preponderance of public universities is being gradually eroded in most Arab countries, one may still notice a strong trend of resistance against privatization in the higher education sector. State monopoly of higher education is seen by many as a necessary political tool to ensure social justice, especially for the poor and the underprivileged in society. This is despite the fact that most public universities are now so heavily subsidized and over-crowded that they have largely become a great burden on society and have not succeeded in delivering the high quality education required by necessity from a higher education institution.

Some critics believe that privatization in H.E. means the commercialization of education an inequality. Public education is based on the ideology of equity. Some argue that despite their weaknesses public universities continue to offer the only source to poor students in societies that remain influenced by issues of class, race and gender.

Traditionally, universities were conceived as experience for an elite few, but, increasingly, they have become a normal part of the educational experience of a larger and more diverse student population. Consequently, this shift has resulted in H.E. more closely resembling a public utility than an experience for privileged elites.

The opponents of privatization argue that it will clearly lead to segregation of students and definitely undermine the common educational experience needed in welfare and democratic societies; that, in addition, the profit motive and the very capitalistic structure of most private universities will, by necessity and definition, lead to lowering academic standards; and that, at best, if private universities are allowed to operate they should do so as non-profit-making institutions, and should be heavily controlled and regulated by the State.

Alternatively, advocates of privatization in higher education, view it as a part of the inevitable globalization trend. They argue that such universities provide better choices for citizens in addition to innovation through competition in the market place. They claim that past experience shows that the market will regulate itself through competition and the quality delivered, and that poor quality education will, by necessity, be unable to survive in the long run.

Proponents of privatization stress the efficiency of a free market rid of bureaucracies. Thus, an "education industry" has evolved all over the world. Global education industry was estimated at a staggering USD 2 trillion (BHD 0.75 trillion) in the 1990s.

A wide variety of companies are taking over most of the work traditionally performed by public universities, from teaching to providing student transportation, to cooking meals to cleaning and maintenance, etc.

In the United State of America, for example, the overwhelming majority of higher education institutions (over 3000) are still State universities and non-profit-making. However, this trend is changing fast. According to Wall Street analysts, private education companies represent one of the fastest growing areas in the US economy.

Today, they account for only around 2% of the USD 211 billion (BHD 79.3 billion) spent annually on higher education in the U.S., but they are expected to grow to between 15% and 20% of the total soon. An estimated 100 private corporations are operating their own institutions of higher education, while others are working in partnership with colleges by making funds available in critically needed educational and professional fields.

Software giants such as Microsoft and Oracle and others need highly skilled workers and specialists. They are now investing considerable sums in specific programmes to ensure that a pool of highly qualified graduates is readily and steadily available for their industry.

Some educationists argue that the incredible growth in private universities run by corporations clearly means that public universities have, more or less, failed to meet the needs of the market place and industry. There is a growing public concern that most public universities, including Arab ones, are clearly failing to meet the intellectual needs of students.

They also argue that the emergence and the relative success of private universities in higher education were due to the fact that they provided more choices to the consumer. They propose that the dynamics of competition should be viewed as a positive step toward improving the overall quality of higher education and thus becoming an important factor in attracting students.

The concept of "privatization" is well developed and elaborated in the economic and industrial sectors, but not in the educational sector. It is not easy to define this concept because it comes in many shapes and forms. It dose not have a clear-cut definition because, contrary to many other concepts, it does not relate only to ownership, finance and control. Privatization in its holistic framework should be viewed as an integrated process within the parameters of a free marketplace and economic structure, including flexibility, profitability, competition and quality.

Therefore, there are many possible shapes and forms, which a 'private' university can take. It dose not relate to who owns it. In India, for example, some public universities assign certain academic programmes to be run by selected private colleges that are completely financed by the state through grants or vouchers given to enrolled students.

In the US, many independent universities are not owned by individuals but run by trusts and do not operate for profit. Similarly, it is not a question of fees because some so-called public universities charge more fees than some private universities.

Some public universities in Jordan, for example, have opted for various means to supplement their revenues. These include so-called 'parallel' programmes and 'international' programmes as well as 'evening studies' programmes. Students enrolled in these programmes pay similar or even higher fees then most private universities in the country.

It may be argued that this is a form of 'privatization' creeping into the public sector. Such public universities are now directly competing with the private sector, although they are reluctant to admit it. It may still be argued that these same universities could obviously undermine the very philosophy and mission of education in public universities. There is nothing wrong with raising the fees for all students who have the means to pay more, but to create double standards of educating and admission should be looked into thoroughly.

Similarly, many pubic universities around the world do not hesitate to contract private sector corporations to run their dormitories, maintenance, and catering facilities, etc. Many private higher education institutions in the US receive public financial assistance and various forms of grants. In Virginia, for example, a Tuition Assistance Grant (TAG) program provides private institutions with USD 2,600 (BHD 980) for each fulltime Virginian student they enrol. It is very interesting to note that the rationale for this incredible assistance is that this payment is much less than it would cost the state to enrol the same student.

A few years ago, there were no higher education institutions regionally accredited 'for profit' in the U.S. Now, a profit-oriented institution such as the University of Phoenix claims more that 48,000 'on-line' students across the country and the world, and has shares in the stock market and 18 campuses. Strayer University raises its financial resources like any corporation; it had 12,000 students and some 12 campuses in 2005.

3 TYPES OF PRIVATE HIGHER EDUCATION

The following is a summary of seven types and forms of privatization in higher education institutions:

1. Non-profit universities run by trusts and financed by fees, grants and various endowments (examples: Georgetown, Harvard, Princeton, etc.);
2. For-profit Universities run by corporations and financed by fees, ventures, the stock market and shareholders (examples: Phoenix, Strayer, Motorola, DeVry Tech., ITT Educational Services (USA), etc.);
3. Private Universities receiving grants from the State for enrolled students (example: Virginia (TAG), USA);

4. Private institutions running certain academic programmes for the State against specified fees or vouchers for each enrolled student (many examples in India);
5. Private universities owned by a wealthy individual or only a few business people (examples in the UAE, Yemen, Jordan; etc.);
6. Private universities owned and run by professional associations and unions (examples in Iraq);
7. State universities running certain (private) programmes for students who are willing to pay much higher fees than regular students (examples in Jordan).

4 CONCLUSIONS

It is clear that privatization has increased the pressure on traditional higher educational institutions to operate more efficiently, to goals set by outside interests, and to market themselves more aggressively.

It is also clear that the concept of privatization in higher education is a fluid one and may take different forms. Nevertheless, it can develop into innovative ways to improve the overall quality of education. Such new developments can create the necessary components of competition and partnership with public sector.

Today, the sharp and distinct lines which divide public and private sectors are increasingly becoming blurred or overlapped. Higher education, whatever its financial or management structure, is being viewed as so vital to national development that it cannot and should not be totally monopolized by the State.

Moreover, provided quality assurance measures and accreditation controls are assumed by the State, there is no valid reason to suggest that private universities run against societal or public interest. Indeed, there is no reason whatsoever why there should not be close co-operation and even partnership between private and public universities. Such partnership has proved its success and effectiveness in other sectors of the economy, so why not apply it to the higher education sector?

With the advent of the 'Information Revolution' and globalization, higher education is undergoing tremendous and fundamental changes. These include the intensive use of remote learning and teaching through the internet, computers simulated and virtual programmes, interactive television, and so on. The concept of the university campus itself will be changed, perhaps even disappear in the not-too-distant future. Similarly, the emergence of the "Learning Model" as against the prevailing 'Teaching Model' will have a major impact on the very structure of academic programmes.

Such eminent and radical changes require major re-thinking of the concept of 'private' *versus* 'public' higher education. New ways and means must be elaborated to overcome this stigma, which has plagued so many thinkers and educationalists for so long.

Higher Education in the Twenty-First Century: Issues and Challenges – Al-Hawaj, Elali & Twizell (eds)
 ISBN 978-0-415-48000-0

Educational outcomes and industry needs

Nabeel Al-Jama
General Manager Training & Career Development, Saudi ARAMCO

ABSTRACT: Consonant with conclusions drawn from the Eighth Saudi Development Plan, this paper presents Saudi Aramco's business challenges in light of the need to improve Saudi educational system outcomes, to achieve a better alignment between industry needs and human resources. Particular attention is paid to current and desired outcomes in engineering education with a view to bringing into focus desired engineering education attributes (hard/soft skills) from an academic and industry point of view. Saudi Aramco's workforce-profile is delineated with special emphasis on the number of engineers in the total workforce as well as the percentage of engineers at various management levels. Data are also presented that shed light on Saudi Aramco's experience with testing and training thousands of high school graduates. The paper concludes by stressing the need for solid partnerships between academia, business, industry, government agencies and NGOs with a view to aligning their efforts to reform educational systems and to improve outcomes to reach internationally competitive educational levels.

Educational outcomes and industry needs

Higher Education in the Twenty-First Century: Issues and Challenges – Al-Hawaj, Elali & Twizell (eds)
© 2008 Taylor & Francis Group, London, ISBN 978-0-415-48000-0

Research in higher education institutions

Syed Arabi Bin Syed Abdullah Idid
Rector of International Islamic University Malaysia (IIUM)

ABSTRACT: Using Malaysia as a paradigm, this paper discusses various initiatives in promoting research among academic staff and also how universities have benefited from responding to various government measures designed to stimulate research. The Malaysian public universities have instituted various initiatives to develop research among their own academic staff. Such measures include encouraging individual or group initiatives in conducting pure or applied research and the Malaysian government has played a big role in encouraging universities to develop research programmes. The author discusses some of the problems encountered by the universities in conducting research in Malaysia that bear relevance to how private universities in the Gulf might foster indigenous research.

Higher Education in the Twenty-First Century: Issues and Challenges – Al-Hawaj, Elali & Twizell (eds)
© 2008 Taylor & Francis Group, London, ISBN 978-0-415-48000-0

Private education in the Arab World: Overcoming the challenges

Marwan R. Kamal
President of Philadelphia University, Amman, Jordan

ABSTRACT: In the last two decades the Arab World has witnessed a meteoric expansion in higher education involving the establishment of tens of new universities and the matriculation of hundreds of thousands of new students. A number of unsolved major challenges have manifested in the context of this rapid period of growth entailing an assessment of:

1. The capacity, diversity and responsiveness to market demand,
2. The quality of education viewed through the prism of its compatibility with production and labour markets,
3. The governance viewed through the prism of its management of higher educational institutions,
4. The economics of higher education,
5. The role of private universities and their relations with public universities,
6. The place of Arab universities on the quality scale of international universities,
7. The potential for development of a modern Arab higher education industry with productive networking and co-operation programmes,
8. The role of Arab universities in the socio-economic development and progress of Arab countries.

After reviewing these challenges, this paper puts forth possible solutions designed to transcend barriers impeding private education in the Arab World.

Papers and abstracts by other delegates

Higher Education in the Twenty-First Century: Issues and Challenges – Al-Hawaj, Elali & Twizell (eds)
© 2008 Taylor & Francis Group, London, ISBN 978-0-415-48000-0

Challenges and new trends in higher education

Alparslan Açikgenç
Vice-Rector, Fatih University, Buyukcekmece, Istanbul, Turkey

1 INTRODUCTION

The university as the home of higher education has always faced challenges in history. We would like to discover the challenges in our time. Since the university is a part of the general education, in order to discover these challenges we must try to understand the nature of education, which can broadly be understood as a "process of acquiring knowledge." For in the actual sense education consists of learning and teaching. The former is acquiring knowledge and the latter is imparting knowledge. In that case both teaching and learning are processes of knowledge. But since the aim of education is learning then our attention must be on this aspect. Therefore, if we analyse the nature of learning as the process of acquiring knowledge we may at the same time discover the real nature of education. This way I am hoping at the same time to develop more effective methods of teaching in education.

The process of acquiring knowledge has two aspects: one is external, the other is internal. What we mean by the internal aspect of the process of acquiring knowledge is the operations of our faculties which we employ while acquiring knowledge. As such the analysis of this process is a theory of knowledge investigated in the branch of philosophy known as *epistemology*. The external process of acquiring knowledge is on the other hand what we call "learning". This is the process which is the subject of examination here as "education". Of course in this external process there are many external factors, such as instructor, teaching methods and other means utilized in teaching and learning, all of which will be briefly discussed in this context.

Learning as an external process of acquiring knowledge can be in two ways: one is natural learning, which is attaining knowledge mainly by the use of our faculties of learning and senses. The most effective method in this kind of learning is trial and error. The other kind of learning is systematic knowledge acquisition. It is this second kind of learning that we should consider as real education. Now we come to a point where we have to recognize the role of epistemology in education because education as systematic learning is an external process acquiring knowledge, which is based on the internal process of acquiring knowledge. For, the internal process of acquiring knowledge as learning is mainly the operations of our mind and other faculties of learning while we try to acquire knowledge. This means if we know how our faculties of learning operate then obviously we can teach the students more effectively. In that case the internal process of acquiring knowledge can give us more effective methods of teaching and thus we need to utilize it. But this aspect is discussed in epistemology, so I will try only to assume a theory of knowledge on the basis of which I shall attempt to develop a theory of education that may guide us in determining the challenges we face today in higher education.

Our explanation so far reveals that a theory of education consists of the internal process of acquiring knowledge on the one hand, and the learning subject (the human being), which represents the external process including teaching on the other. The former includes an analysis of the internal processes which take place in the mind and learning faculties of the learning subject (the human being). The latter includes the efforts of systematic learning and teaching. Let us then try to analyse these structures in order to pinpoint the place of higher learning in education so that we can discover our challenges and new prospects.

2 THE INTERNAL PROCESS OF LEARNING

The internal process of learning is the operation of our faculties which we use in order to acquire knowledge. These operations begin at the level of objects of knowledge and ends at the level where their complete concepts, ideas and knowledge are transferred to the mind. This process can be natural or systematic. We shall try to understand these processes that will give us an effective method of teaching to be utilized in any education.

2.1 *Natural learning*

The natural process of learning or of acquisition of knowledge is the personal trial of an individual in acquiring knowledge. Usually this is the way we first begin to learn things when we arrive to this world. The main method of natural learning is trial and error. As a result of this process a worldview is formed in the mind of the learning subject. Initially this worldview is simple and as such has only two structures: Life Structure and World Structure. The former starts to be formed as soon as we are born. In this first experience of life it is extremely hard to determine what will be the first that we learn. In other words, what is our first experience that is converted into a piece of knowledge? We can, on the other hand, identify more or less what kind of knowledge this mental content is; for example, it cannot be an abstract idea, or a philosophical notion and the like. It will be a kind of knowledge that pertains to our life at that time. Since we are naturally inclined to preserve our life, most of our experiences will be related to the preservation of life, such as finding and choosing certain food and developing habits of how to attract the attention of others to make food available and so on.

Therefore, in our early ages we *naturally* have such experiences available to our mental consciousness and it is these experiences that are converted into knowledge. What we call Life Structure is the knowledge that is available for us in this way and is primarily related to our biological and daily life. In that case Life Structure in our worldview includes most of our daily habits related to the preservation of our life. This Life Structure becomes more refined and sophisticated as we add to it what we learn from our social environment. In this way it begins to include many of our cultural habits as well, such as habits of eating, our social culture, and the ways of daily behaviour, manners and customs. We call it 'structure' because our mind forms it according to its natural rules and principles explained in such disciplines as logic and epistemology. For this reason, knowledge gathered in this structure is not a hodgepodge gathering of experience in our mind; it is rather an orderly unity according to certain rules and principles. Life Structure is, therefore, such a coherent mental unity which makes up the total contents of our mind in our earliest life onwards enriching itself until adulthood according to the natural rules and principles of the mind through its social and physical surrounding. It is this internal process that we call natural. When there is inter-ference from either us or our surroundings then learning becomes systematic.

I would like to illustrate the natural internal process of learning with a concrete representation. We have a digestive system which is totally biological. Yet, if our capacity to attain knowledge is called "knowledge system" following our naming of the "digestive system", then it actually resembles our "knowledge system". Our digestive system, too, has internal and external processes, both of which also have "natural" and "systematic" ways, although they may not be called as such in anatomy. The former is the process which takes place after the food is put into the mouth; and the latter is the process which takes place during the gathering and preparing the food before it is eaten. Hence, the internal process of digestion is like the internal process of learning or of acquiring knowledge, namely the process that takes place after the "information" received from the objects of knowledge is put into the *mind*. The external process of digestion is a good illustration to the external process of learning or of acquiring knowledge, namely the process that takes place before the "information" received from the objects of knowledge is put into the *mind*. It is clear that in all these processes there are natural and systematic procedures. We have explained so far the natural internal process of acquiring knowledge. Now let us see the systematic learning or acquiring knowledge.

2.2 *Systematic learning*

As far as the internal process of learning is concerned there is no difference in natural or systematic learning. Here, the difference comes because of the external applications of our learning faculties and as a result in the mind a more systematic worldview is formed. This makes a significant change in learning. For, as soon as we begin to form a mental conception of a natural experience, which we had from babyhood onwards, we will begin to act no longer out of the natural instincts alone, but also out of the mental content that we have acquired which we have called 'Life Structure'. The more sophisticated the Life Structure is the more conceptual becomes the experience and thus the more we act out of our mental frameworks. This means we are now acting on our knowledge which we have thus far acquired and hence knowledge formed thereupon is more systematic. In such a conceptual Life Structure we may be able to distinguish certain elements, which we call 'mentality'. A mentality is actually an understanding or conception of certain things, living types, and facts of life and of the world. As we grow these mentalities are developed according to our personality, mental abilities and the kind of education we receive. Each mentality begins to develop through the education we receive into a structure and thus can be termed 'sub-structure'. These mentalities are so coherently related to each other that together they form the totality of the Life Structure. Then, we begin to arrange our lives according to our own Life Structure, which is the totality of the contents of our mind thus far. Since, as a total unity, the mind reflects all the ideas we have thus far formed in our mind, its contents as the Life Structure will also reflect our attitude for life and understanding the universe in general. Our worldview at this stage has only a Life Structure which reflects our conception of the universe, such as the meaning of life, the origin of existence, human destiny and so on.

As we grow up certain conceptions concerning the world we live in gradually develop in our worldview; first, certain fundamental questions arise in the mind, such as the meaning of life, from where we have come and where we are going to. As we try to answer, or find answers to these fundamental questions, a conception concerning the world and things around us is formed. As this conception begins to be more sophisticated through our education, it gradually forms a clearly discernible structure in the mind, which can be distinguished from the Life Structure. This new structure first forms a mentality within the Life Structure. But, if it can be fully developed, it provides well-formed ideas, doctrines and even scientific opinion concerning the world we live in; hence, it is apt to call it 'World Structure'. As soon as this new structure is established within the worldview, it begins to function in conjunction with the life-structure and *vice versa*.

If there is a good education in the community then the worldview in our mind develops further. Usually there are certain concepts in our worldview that dominate our lives. It is possible to reduce these concepts into only five fundamental ones: life, world, knowledge, human and value. A fundamental concept is one which forms a complicated understanding and a mentality in our mind which directs us to a certain behaviour in life and society. As such, these concepts can be characterized as "doctrinal concepts" because each one of them may constitute a specific well-developed doctrine which determines our worldview. The development of these concepts depends upon the type of education a person receives. This brings forth the importance of education which will be evaluated as the external process of acquiring knowledge. We should be able to see here that at least two doctrinal concepts grow into structures in our worldview as a natural internal process of learning; Life and World Structures. But if the education is scientific then the concept of knowledge can also acquire a doctrinal characteristic and as such formed into a "Knowledge Structure". In the same way both human and value concepts can also develop into doctrinal concepts and as a result form the structures of Man and Value. What is important for an educational philosophy is the process in this development. Two structures, namely, Life and World Structures, develop as a result of the natural internal process of learning or of acquiring knowledge. We must, however, realize that no structure is perfect in relation to knowledge unless it is developed through systematic process of learning. That is why the other three structures, Knowledge, Human and Value, can develop only if there is a systematic process of learning. Therefore we shall build our educational philosophy on the system of the development of a worldview. Before we do this, however, it is important to clarify further our concepts of structures and the way they lead us to certain behaviour in life. In fact we need to prove that it is these concepts in our worldview that guide us in our actions.

First of all, we need to show what other concepts are restored in these structures. The Knowledge Structure includes within itself the key scientific terminology and all other concepts of technology that may be derived from these concepts. The network of the key scientific terminology may be called "Scientific Conceptual Scheme', which will be dealt with briefly below. In the Value Structure we may find moral concepts, ideas, doctrines, and depending on the kind of worldview, we may also have our religious and legal conceptions. In the Human Structure, on the other hand, we have our conceptions of ourselves, as well as of the society and the societal organization. All structures of a worldview operate in relation to each other. None of them can operate independently; hence, our treatment of them independently is only a logical analysis of a worldview. Otherwise, it is not intended to establish each structure independently.

Secondly we may give a concrete example in order to illustrate the structures of a worldview on the basis of the Islamic worldview. Since the Life Structure is grounded in human biology, it will have the most common elements with all other worldviews, and as such the Life Structure of the Islamic worldview is its aspect that is most dominant in the Muslim local cultural activities. The World Structure is that aspect of the Islamic worldview which includes the most fundamental elements, such as the idea of *tawhîd*, (God's Being and Oneness), prophethood, resurrection and the ideas of religion and the hereafter, *akhirah*. We do not mean that these are the only fundamental concepts of the Islamic worldview because each structure by itself represents a doctrinal element which includes within itself many other fundamental Islamic key terminologies. But the extensions of these key concepts and terminology constitute substructures; hence, there lie many substructures within the basic structures of the Islamic worldview which may not be so fundamental and as a result differences of opinion in those sub-structural elements can be allowed. As an extension of the World Structure, Knowledge Structure is also a fundamental doctrinal element, which is represented by the umbrella term *'ilm*. This structure includes within itself the key scientific terminology of Islamic science which may be called 'Islamic scientific conceptual scheme'. The Value Structure in the Islamic worldview, on the other hand, includes moral, ethical and legal practices. But since the concept of law; namely, *al-fiqh*, in the early Islamic worldview is closely linked with the World Structure, it naturally included religious law, which cannot be devoid of moral content. Hence, law, religion and morality are manifested as an integral part of one structure. This conceptual understanding of law, religion and morality never brought about a sharp distinction between these three sub-structures. Finally, the Human Structure is represented within the Islamic worldview by the concepts of *khalîfah* and *ummah*. As such, this structure manifests the Islamic understanding of man and society, which is totally grounded in the World Structure because, again, even these conceptions themselves are derived from the concepts of *tawhîd*, prophethood, religion and *âkhirah*.

Thirdly, each structure in a worldview has a specific function in life and in human activities. This point can be explained from another perspective as well; let us assume a worldview in which the Knowledge Structure is not discernible as a manifest mentality. In such a case, no scientific activity will ensue from the individual having such a worldview. There will not be in that worldview any scientific concepts that can form a scientific framework for the mind to work in. As a result, there will be no scientific attitude, nor any scientific tradition that can support such activities. In fact, if there is no Knowledge Structure within a worldview, then that worldview can only be analysed into its Life and World Structures. For it is the scientific activity which manifests other structures as analysable units of a worldview; if there is no such activity those structures cannot be developed to such an extent that they become manifest in their respective worldviews. This does not mean that a worldview without a manifest Knowledge Structure lacks a value system, or a Human Structure that acts as the ground of social and political activities; on the contrary, all these activities will be carried out and regulated by a World Structure that may acquire a degree of sophistication within its respective worldview. But it cannot acquire the level of sophistication manifested in such scientific worldviews that can adequately be analysed into their Knowledge, Value and Human Structures. This is where the significance of what I am inclined to call "scientific worldviews" lies. For, it is only these worldviews that can clearly be analysed into their manifest structures.

2.3 *The operations of the internal process of learning and the faculties*

The operations of the internal process in education constitute a fully-fledged epistemological theory. The operation itself depends on our faculties of knowledge. Therefore, we need to analyse this process as an epistemological function of our mental faculties which shall reveal both the internal learning process and the faculties at the same time. In the first place this process begins at the level where our faculties of external experience somehow get in touch with an object of knowledge. Here we may ask: which faculty is it that gets in touch with such an object? This depends on the object; if it is a material object then it is received into the mind by way of one or more of the five senses called external senses. If the object of knowledge is non-material then it is received into the mind through one or more internal faculties. There may be objects whose nature is such that we can never get in touch because we lack the faculty to get in touch with it. Such objects may exist but remain unknown to us and therefore it is futile to attempt to acquire their knowledge and to convey such knowledge. This tells us also where education may begin and where it should end.

When an object of knowledge is thus received into the mind the impression we form is called "mental representation" which is the result of an object's undergoing through the process of experience. Since there only two categories of objects for knowledge, there are also two kinds of experience, external and internal, both of which take place differently. The latter depends on the former; therefore, external experience develops in us first, and thus it is in the first place, the beginning of the process of experience, and in the second place, the beginning of the whole process of knowing. Any philosophy of education must pay attention to these processes in order to form a theory of education.

The awareness of an object of knowledge produced in us is perception. If this awareness is received through one or more of the five senses, then it is a *sensible perception*; this means that whatever is received through the five senses can be qualified as 'sensible', and everything that is received in this way can also be termed 'sense datum'. All the data of the sensible perception are what we call 'external' or 'outer experience'. Since we have no senses other than the five senses to be in contact directly with the material reality, the ultimate beginning point of the process of acquiring knowledge concerning the physical universe is sensible perception, which originates from the five senses. The process of outer experience begins at one or more sense organ(s) and ends at the *physical consciousness*. Since this process as a whole is totally physical, it needs mental consciousness to be conceptualized. If a sensible perception, which is yielded at the physical consciousness, is not conceptualized it cannot be utilized as an object of knowledge. Therefore, such a perception needs to be perceived by the mind as well so that it can be a proper object of human knowledge. This leads us to posit the existence of a mental perception on the one hand, and of a mental awareness on the other. So far the internal process of knowledge has been explained to the point of the faculty of physical consciousness where all sensible perceptions are produced. These perceptions then are utilized in two ways: they are directly perceived by the mind and thus are on the way to be conceptualized; and/or they are taken by the faculties of internal experience to be utilized, to use Kant's term, as raw material in their functions. This means that the process of acquiring knowledge after the physical consciousness will continue to two directions: to the direction of the internal experience; and/or to the direction of the mind. But since the product of the internal experience is also utilized by the mind in the process of knowing, we shall first try to first analyse the direction of the internal experience, and then continue to the analysis of the mind. The *inner* perception of an object of knowledge is an emotion. The fact that emotions play a role in the process of acquiring knowledge cannot be denied. Therefore, the role of emotions cannot be denied in education either. For, an emotion moves us, excites us or makes us feel miserable, but how can it contribute to the process of acquiring knowledge?

We may answer this question by illustrating with the example of a lion. Suppose we do not have any feeling of fear; would we still run away from the lion even if we knew that a lion is a formidable and dangerous animal? I am not saying that we still have the emotion but we simply do not fear anything; on the contrary, I mean we have absolutely no such emotion at all. I think it is the emotion and not the sensible perception of the lion that is mentally evaluated and then decided whether to run away from the lion. For sometimes we think that we have seen a lion, although there is no lion and we are immediately activated to find a shelter. Similarly the idea of a lion by itself is not sufficient to make us look for a shelter, even if it is sufficient by itself, it will not prompt us immediately to move. Hence, the decision to look for a shelter is considerably influenced by fear. We would like to formulate this conclusion in the epistemological sense as 'the internal perception of an object of knowledge is given to the mind as raw material' just like the sensible perception of an object of knowledge. In other words, our mental faculties of knowledge cannot function without our faculty of outer experience, nor can they work in isolation from the faculties of inner experience.

The internal experience which is nearest to the sensible perception is the *sensible intuition* which may perceive an object without physically perceiving it, because no physical object can directly affect our faculties of internal experience. Therefore, if an object is intuited as existing this is the result of the function of the sensible intuition. We call this faculty 'sensible intuition' because it intuits its object the way our mind intuits the conclusion of a deductive argument, and yet the intuition of an object is not mental intuition; since, though a sensible perception is absent, yet it is assumed as mediating between the faculty and its object, it is apt to call it 'sensible intuition'. It is the faculty of sensible intuition that produces our instincts, which are direct and immediate; and as such they can play a significant role in the manifestation of an emotion. Because of this function, the sensible intuition may interfere to mediate between the sensible perception and the other faculties of inner experience. In a sense, it intuits the data of the sensible perception and makes them available for the other inner faculties. When such a sensible intuition of a sensible perception is made available for an emotional state the corresponding emotion may be yielded. Another important function of the sensible intuition is that it retains the copies of our emotions through which we instinctively recognize such emotive experiences, because it is the centre of our instincts.

The central faculty of our inner experience is *heart* which assesses our emotions in the sense that it renders them meaningful for the ethical and religious life. The function of the heart in the epistemological process is to provide our mind with representations of ethical and fundamental concepts of life and death in an experiential way. Heart is not a faculty of outer experience and as such it cannot perceive external objects, but it can perceive what we may call "emotive objects of knowledge" which transcend our external experience. The sphere of knowledge comprehensive of all things that may be subjected to the emotive process may be identified as the 'Experiential Realm' which is transcendent to the external experience. With this conclusion it is clear that we are distinguishing two comprehensive realms of knowledge: the Realm of External experience and the Realm of Internal Experience which we have identified as Experiential Realm. All these realms are to be taken into consideration in educational curricula as well. On the other hand, the Experiential Realm has also two further spheres; one moral, the other religious. The former is clear in the sense that its limits can be quite easily drawn, but the latter needs further clarification. In the theological sense, both realms involve the same subject but different aspects of it; whereas the experiential realm, for instance, encompasses God as an existent Being, the Transcendent Realm has God in the Absolute sense, namely as He is in Himself. Therefore, the experiential representation of God's existence is an inner experience, but His Essence is an Absolute behind "seventy thousand veils" as expressed in a tradition of the Prophet. Moreover, our mind also needs the guidance of the Revelation in attaining the theological experiential realm. For as we shall see, our intellect operates with causality in order to infer a Maker for the whole existence; it may thus discover a *Cause* for the universe, but it still needs another argument to prove that that *Cause* is God as defined in the Revealed Texts. Is the experiential perception provided by the heart sufficient to make such an inference? This is the greatest dilemma of philosophy and it seems that it is still not sufficient and therefore we need a *guidance*, which is provided by the Revelation. This epistemological principle guides us as to which method is to be used in education of these subjects.

Experiential inquiries belong directly to the inquirer, though they are not perceived directly. What I mean by this is that our inner faculty of representation needs an awakening by and the mediation of Revelation. Just as our faculty of outer experience sometimes needs the mediation of a device to see or hear things afar, though they are perceivable directly; in the same manner the inner faculty needs a finer and more subtle mediation. Again this mediation is guidance, which is reflected in the mental states of the inquirer, and these states represent the terms of the internal experience. Since as such it is primarily our emotions that yield such a mental state we would like to call it "subjective mood". As a mental state, the subjective mood is very important for the faculties of inner experience to perceive the truths of the experiential realm. Our analysis of the process of knowledge has brought us to the faculty of ethical feelings, called 'conscience'. Therefore, heart does not represent all our internal faculties, although it is the most fundamental one, because it is through the heart that most of our moral and religious experiential representations of entities as objects of knowledge originate. This approach solves an issue that remains inadequately answered in the Critical philosophy of Kant.

This process of knowledge through internal experience converges into and culminates at a unity which makes us to be aware of our 'self', and as such it functions as a faculty, which we term 'the faculty of consciousness' or 'faculty of the self'. Through this faculty one experientially becomes aware of his/her existence which is very similar to the external experience through which one becomes aware of an object around him; just as this external awareness is different from the mental awareness, in the same way this internal awareness of the *subject* is different from the mental awareness of the self. Hence, in addition to the external and internal awareness, we must have a mental awareness; the first process of consciousness yields the awareness of things and the world around us, whereas the second one, through the aid of our internal faculties and the experience resulting from them, yields the awareness of our inner world which revolves around the consciousness of our self (ego); but the third one, namely the mental awareness, yields a conception of both the external and internal awareness. Therefore, in order to grasp these experiences in the way a human conceives them, the mental awareness is essential.

All the representations of objects of knowledge are thus transferred to the mind to undergo mental operations. We use the term 'mind' (*dhihn*) in general to mean the 'faculty of conceptual experience'. In this sense 'mind' refers to the totality of our mental activities that are mainly conceptual. Such a conceptual understanding, as we have seen, begins at the level of mental consciousness, which is thus the lowest faculty of the mind. In order to discover all mental faculties we need to examine the functions of our mind. Let us try to show the internal process of learning thus far outlined on a table (Table 1).

Table 1

FIVE SENSES →

PHYSICAL CONSCIOUSNESS
Sensible perception is produced
which can also be called sense data

EXTERNAL EXPERIENCE

INTERNAL EXPERIENCE

HEART →

SENSIBLE INTUITION
THE SEAT OF EMOTIONS
CONSCIENCE
INNER CONSCIOUSNESS

MIND

First, the mind must somehow retain all conceptual activities so that it can utilize them in all its functions and this leads us to the faculty of *memory*; second, objects of knowledge must be presented to the mind in a conceptually concrete way, which means the reproduction of representations as images and this poses a *faculty of imagination* (*khayâl*); third, the mind must think the objects of knowledge, which leads us to postulate a *faculty of thinking* which we call 'intellect' (*'aql*); fourth, since the mind as a result of thinking must decide in order to know its objects, and that deciding is actually an act of 'choice', the mind must need a *faculty of judgment*, which is already known as the 'will' (*irâdah*); and finally, the mind functions to formulate arguments and reach conclusions out of these arguments, which means that it must somehow possess a faculty which enables it to deduce the implications and entailments of organized propositions, and this is the faculty of inference which we would like to call '(mental) intuition' (*hads*).

All these mental activities are already known to us; they are not new discoveries that must be studied now. In fact, we apply them all in our daily life without thinking that this process which has been taking place in our mind can be analysed in this way. We do not have to know the process of knowledge in order to acquire knowledge; just as we do not need to know how our stomach functions in order to digest the food we eat. For all the functions of the mind are already provided for the process of knowledge which is, therefore, a natural process but we may interfere at certain levels of this process, which makes it different in this sense from digestion. Hence, we do not need to know this process in order to acquire knowledge; but we need to know it for scientific reasons; for instance, if some one argues that science is a universal and absolute activity which cannot change from culture to culture, then, in order to show that this is not the case we must explain how

Table 2

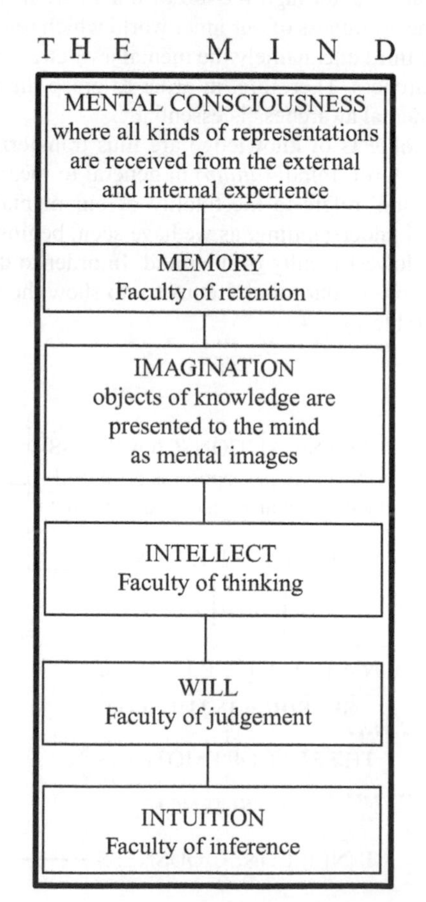

we acquire knowledge in general and then scientific knowledge in particular. Hence, in this case we need a theory which explains how we acquire knowledge.

For educational purposes also a theory of knowledge would be very useful, because if we know how humans acquire knowledge then we can teach them with more effective methods, since teaching primarily consists of "making the student acquire knowledge", i.e., 'to learn'. These mental operations can be illustrated in Table 2.

That activity of the mind which holds the object of knowledge in abstraction is called 'thought'; the faculty through which thought is actualized is the intellect. Thought or the activity of thinking itself does not necessarily include judging and inferring, although ordinarily we refer to all these activities as 'thinking'. In our analysis of the process of acquiring knowledge, we would like to distinguish the thinking activity of the mind which involves judgment and inference from the kind of thinking which is a mere reflection on the object of knowledge and as such requires only the activity of the intellect, which can thus be called also 'intellection'. The intellect is also able to develop certain principles. This is done in two ways: the mind may apply its own natural principles, such as contradiction and identity to its own content and derive new principles, or it may use the natural operations of the intellect to derive a principle; such as the principle of causality.

Since the mental intuition is the last faculty in the process we may confidently say that our analysis of the internal process of learning ends in this faculty. We may show our result in Table 3. This process does not always repeat itself in the same manner. This is because our faculties of knowledge can make the use of previously available data, whether from external or internal experience, or yet from the sources of our mind itself. So, the actual operations are more complicated

Table 3

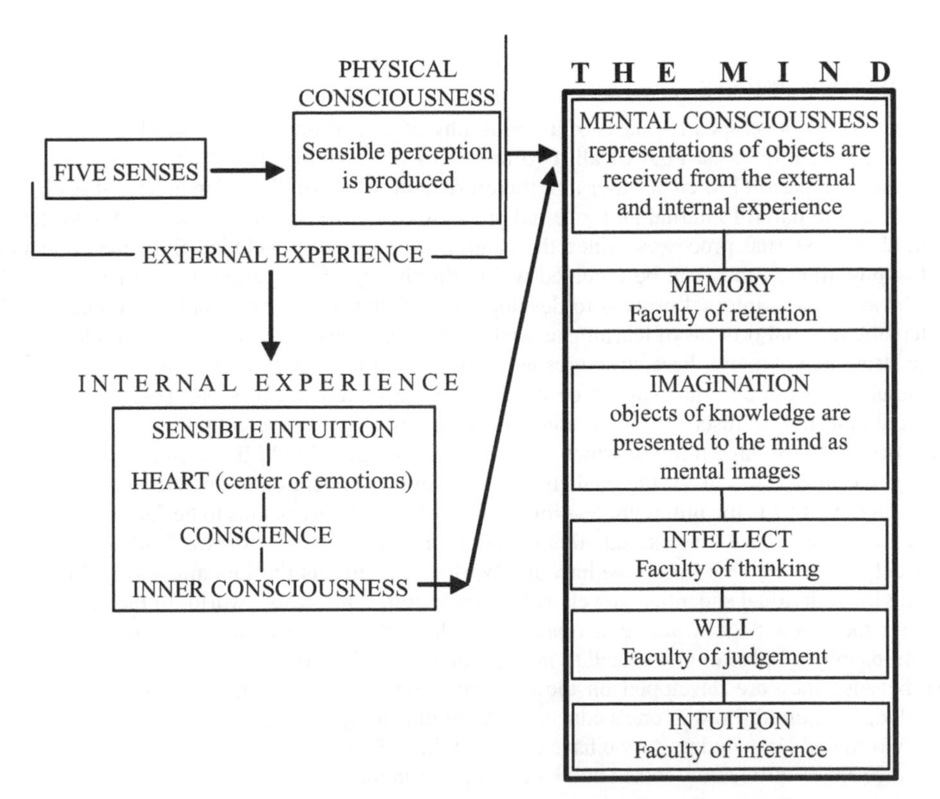

than this systematic exposition. However, we may utilize this exposition to establish a philosophy of education on the basis of which we can erect our theory of education.

3 THE EXTERNAL PROCESS OF LEARNING: PHILOSOPHY OF EDUCATION

The external process of learning is all the activities taking place outside the learning subject when someone is engaged in learning and hence, it is education in the real sense. If we fully outline this process it will be in the true sense our philosophy of education on the basis of which our educational theory will be developed. We are not in a position to fully outline it in this context because there are more concepts involved in it, such as the purpose of education and the nature of the subjects to be educated. We shall therefore concentrate on the main problems that will guide us through the challenges of higher education. Since the internal and external processes do not work independently of each other during the actual activity of learning it should also take place at two planes; natural and systematic.

3.1 *Natural learning*

As we have pointed out, the natural process of learning or of acquiring knowledge is the personal trial of an individual in acquiring knowledge. Usually this is the way we first begin to learn things when we arrive in this world but it continues after even we begin to learn systematically through our regular school education. Everyone can develop his/her own method for this kind of learning but in any case the main method of natural learning is trial and error. Because of this individual character of natural learning no educational theory can be established for it. The actual education evolves when systematic learning begins in a scientific way and we shall now examine this.

3.2 *Systematic learning*

We have so far developed the basis of a philosophy of education. Now we need to eradicate our philosophy on this basis. First of all, two processes of learning, namely internal and external processes are parallel to each other and therefore cannot be totally independent of each other. If so, then the natural internal and external processes are also parallel as well as the systematic internal and external processes. Since the natural processes are excluded from the educational philosophy, though they can be included within the theory of education itself, then we need to use the systematic internal process to develop the skeleton of our philosophy of education. The systematic internal process of learning as we have seen gave us the concept of worldview with its basic structures. Among these structures as we have seen the Life Structure is developed within the natural process and thus cannot be included in the systematic process. This does not mean that it should not be discussed in an educational philosophy. On the contrary, the epistemological process at this stage through which a worldview emerges should be discussed here. But it is not significantly relevant for the challenges facing higher education today. Hence we suffice by indicating that this is the initial process through which worldview begins to be formed in the mind of an individual. If we thus exclude this structure then we are left with two fundamental groups of structures: in the first group we have the World Structure, as the fundamental outlook representing the individual's identity as well as his perception of the whole world. In the second group we have the Knowledge structure as representing the rest of the structures. We argue here that in the philosophy of education we need to include the rest of the structures in the Knowledge Structure because they are developed on the basis of knowledge and indeed through the scientific knowledge acquired through one's education. With this analysis we are left with three structures: Life, World and Knowledge. As we have excluded the Life Structure from the philosophy of education we are to utilize only World and Knowledge Structures.

Since the Knowledge Structure regulates our scientific activities it also includes within itself the network of our scientific terminology, which we have called "Scientific Conceptual Scheme". As this is a developed mentality it must be included in the higher stages of education. It is therefore inevitable to include the Scientific Conceptual Scheme in our philosophy of education. This outline gives us three stages in education:

1. The Stage of the World Structure and as such it constitutes the earliest systematic learning process. Since at this stage the worldview of the individual is not fully developed to be distinguished from his/her World Structure it is more apt to call this the "Stage of Worldview".
2. The Stage of the Knowledge Structure which constitutes the middle stage and as such the basic terminology of sciences are given and the Knowledge Structure of the worldview is fully inculcated into the minds of individuals.
3. The Scientific Stage where attention is paid to develop in the worldview of the individuals a scientific mentality. Since this represents the final stage of education it must be the stage of higher education.

We can add another stage to this philosophy of education which is the Stage of Specialization. This stage develops a more specific network of concepts in the minds of individuals called the "Specific Scientific Conceptual Scheme" in which the nomenclature of individual disciplines is harmonized.

It is possible to dwell upon these stages in more detail. But because of space and time limitations I would like to skip this detail and rather concentrate on deriving an educational theory from this philosophy of education. Our educational theory, as based on this philosophy of education, reveals five stages of learning:

1. The early education which has the Life Structure as the most important component and starts from the birth of an individual before the formal schooling starts. But we must add that this education should continue throughout one's life because it concerns the Life Structure which is also the initial worldview.
2. Elementary Education which is concerned with the worldview of the individual.
3. Middle Education is concerned with the Knowledge Structure within the worldview of the individual.
4. Higher Education is concerned with developing the Scientific Conceptual Scheme within the Knowledge Structure, and finally
5. Graduate Education aims at developing the Specific Scientific Conceptual Scheme and as such it is the stage where more attention is paid to the specialization.

The last two stages concern us here as they deal with higher education. But we need to elaborate all the stages in order to clarify how one comes to this stage. First of all in the early education the individual is prepared for the early education and taught the cultural elements in which knowledge should occupy a special place. This way every person knows that learning is significant. At this stage the home environment and what the parents do is very important. An individual with a good Life Structure in mind means one who has a worldview that backs up knowledge tendency. I believe that it is at this stage that the person is motivated to do what he/she wants to do later in life and as such it determines for the most part the person's tendency to develop though somewhat unconsciously and naturally his/her leaning for a career in life. It may play a role to develop the person's inherent ability for whatever it may be.

Secondly, the individual is ready for the Elementary Education when he/she has a solid Life Structure in mind. Since this corresponds to the Stage of Worldview at this stage of education only elements that make up the basic structure of the individual's worldview must be given. Special techniques should be developed to the teaching of his/her worldview. The curriculum should also be developed on the basis of that worldview. There is no need to teach basic sciences at this stage as is done today in the Muslim world.

In the Third Stage the Knowledge Structure of the individual's worldview is given. We need to elaborate this because it is crucial for an educational theory. Also it is this stage that prepares the student for higher education. In the Knowledge Structure of a worldview there are concepts that provide a mentality to the individual. These concepts are held in unity under the umbrella concept "knowledge" which thus acquires a doctrinal character. Other related concepts in this unity are science, truth, falsehood, opinion, belief, certainty, method, theory, understanding, doubt and so on. All these concepts are well formed and harmonized together so well that their unity projects an understanding called "knowledge mentality". As such it projects one's understanding of knowledge, method and truth together with their significance. The student learns with this mentality that knowledge is valuable and that one needs knowledge in life. Also the student finds out that knowledge acquired with special method is called "science". The application of science is technology and so on. Of course the way these propositions are put here may be expressed in different ways in every worldview. There is just one common characteristic in all worldviews here; they all have a specific Knowledge Structure of their own. When this structure is well developed in the minds of the students then they find out what to do with knowledge. Those who are still interested in the same type of systematic knowledge will continue for higher education where they will begin to learn all sciences in a general way. Hence, we arrive at the Fourth Stage which is the stage of Higher Education. The approach at this stage should be interdisciplinary so that the student learns about all the sciences in a general way but in later years of this stage they drift to a more particular area which will eventually become their area of expertise.

After the student graduates from this level which corresponds to what we call today the bachelors degree he/she continues for deeper specialization in graduate studies which represents the

Table 4

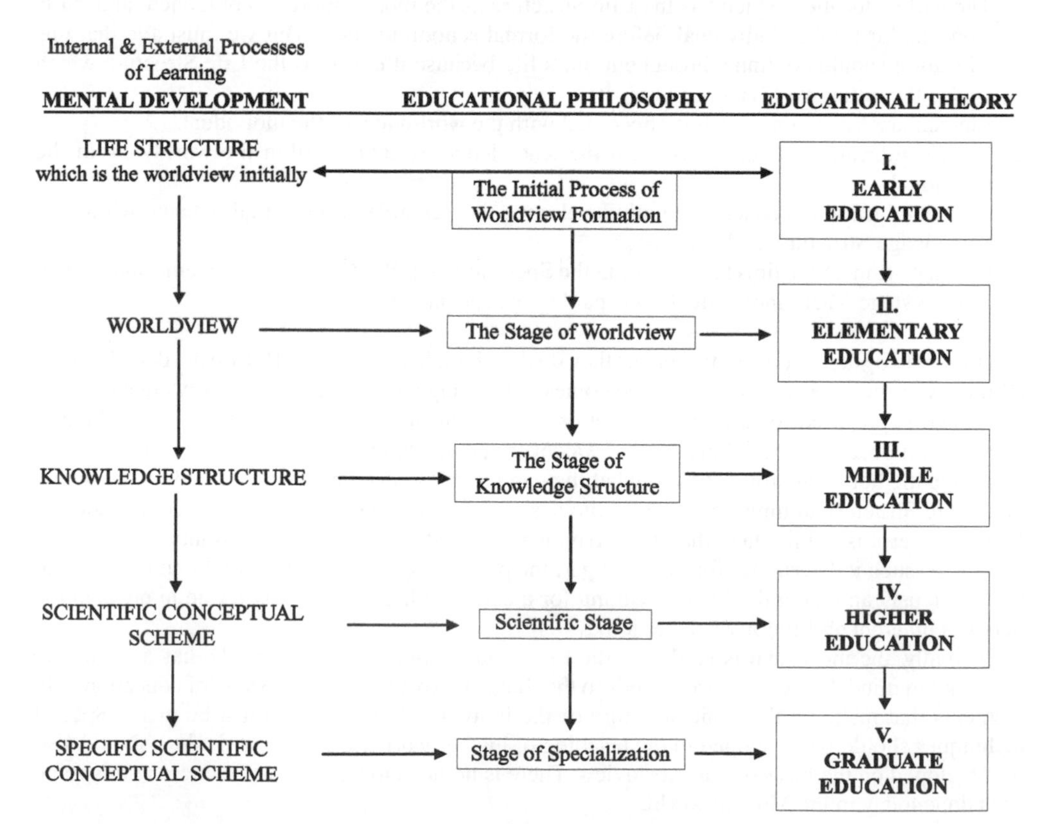

Fifth and final stage of our educational theory. We may now summarize the whole educational process in Table 4.

4 CHALLENGES FOR HIGHER EDUCATION

Based on our exposition above the most crucial challenge facing the university in the Muslim world is what I call the *discrete approach*. This approach takes higher education as a separate entity from the early education and does not consider it as integral and complimentary to it. In the past the *madrasa* education was not taken discretely, but as a whole from the beginning to the end of one's educational training. But today the university is not seen any more a part of this general education but rather the training sanctuary for either practitioners of science or for those who seek to get a job for livelihood. As a result of this wrong understanding most students come to the university not to acquire the scientific mentality of the age. They rather come with a wrong attitude thinking that they need to acquire the contemporary scientific knowledge. Above all most students think that acquiring this knowledge will give them a better chance to get a good job. They never think about how to be more successful after getting the job. As I see these problems more related to the internal process I would like to evaluate them as "Internal Challenges".

Challenges that face our higher education today are not limited to the internal problems of education. There are more challenges that face us outside the educational field also. These are political, economical and moral. Since these problems do not concern education directly and hence external to it I would like to evaluate them as "External Challenges".

4.1 *Internal challenges*

The attitude dominating the internal challenges is the discrete approach which leads to two more misunderstanding. The first one is the usual practice which sees the university as an institution of teaching and thus only scientific knowledge is conveyed to the student but scientific attitude is not taught. Scientific attitude is a *skill* which is hard to acquire. But acquiring knowledge is not a skill and therefore, knowledge can be acquired without even having such a scientific attitude. We need to elaborate on this attitude to show its significance.

Scientific attitude is a skill that is projected with a mentality that develops in a scientific tradition. A mentality is an understanding that is grounded in a structure of a worldview. As such it controls our behaviour. It is in a sense the mental framework or perspective out of which *naturally* and/or *actively* follows a human activity. Knowledge proceeds from a mental framework *naturally*, if it arises purely out of the capacities of our knowledge faculties. Therefore, if an activity springs only naturally from a mentality, then it depends totally on the internal process of knowledge which we have outlined above. But usually humans are not subjects that acquire knowledge passively; they are rather active agents who contribute to the process of knowing and learning. In this way knowledge will proceed from the knowledge acquired through both our education and using the natural capacities of the mind. It is this kind of a knowledge acquisition process that we call 'active'. It is clear that mentalities are more important in the active knowledge processes because it is the mentality that controls the active participation in knowledge.

After this brief outline of mentalities we may show how important it is to develop a scientific mentality. For, without such a mentality it is not possible to be actively engaged in scientific activities. In order to clarify this, we may give the following analogy: a student who cheats may be said to be cheating because he is selfish, dishonest, and because of the circumstances which led him to that undesirable action. We consider all these and similar motives or circumstances underlying the act to be the *external factors*, because they are not factors to be found directly in the actor and hence can be observed either directly or indirectly within the action itself. Yet there are also certain other mental conditions that lead the student in question to his action, such as his conception of cheating and the placement of that concept within his worldview. This is indeed the mentality which the student has in his worldview. Upon analysing that action, therefore, it is possible to distinguish three elements constituting the performance of it:

1. The mental framework within which the action is conceptualized prior to its performance.
2. Certain physiological and environmental conditions leading to the action, and we would like to broaden the scope of these elements by not limiting them just to the environmental and physiological conditions.
3. The performance of the action itself.

As it is seen in our analogy, the development of these *factors* in the individual's life must take place in different ways, but in relation to each other. The mentality, for instance, does not develop instantly, though the action itself is performed in an instant. Even the environmental and physiological conditions may develop instantly, yet in relation to the mentality and the performance of the action. In fact, the mentality is the framework which includes the totality of concepts and aspirations developed by the individual throughout his life, and as such it constitutes a structural characteristic within a worldview. Since every related concept and event is evaluated within a certain mentality before a decision is taken to perform the action, then it must be considered as the *direct* factor of an action. Since we take worldview to be the *foundation* of mentalities then the worldview itself is the *priormost foundation* of any action. We conclude from this that every human action, including scientific activity, is ultimately traceable to its mentality through its worldview; and as such it is reducible directly to that scientific mentality and ultimately to that worldview. But unfortunately this foundation of human conduct cannot be perceived by observation directly and as a result, it can be overviewed easily; in order to emphasize this fact we may call a worldview the 'non-observable foundation', or the '(conceptual) environment' of human actions. Our exposition of the concept of worldview thus brings us to the conclusion that science arises within certain worldviews only. This conclusion has other implications as well, besides clearly pointing to the fact that in certain worldviews no science can arise.

The proper environment for the rise of sciences means only the adequate worldview within which there is a possibility for the flourishing of sciences. Such a worldview is the one in which, first of all, a sophisticated Knowledge Structure has emerged. Then, as a result of this, a sophisticated network of key scientific terms, which we call 'scientific conceptual scheme', is established by the early scholars of that society in which that worldview predominates. This leads us to conclude that although our scientific activities ultimately derive from our worldview, they do not directly follow out of it. For, there is a need for another framework which directly supports such activities. It is this framework which we may term "scientific mentality". This mentality leads to the development of a skill that is reflected on the scientific activities. Skills are developed through constant practice under the guidance of masters. This can be given only in the university. Therefore this should be the aim of university education and it must have no other concern. We should not worry whether the student does not get the sufficient knowledge which he will need after graduation when he starts working in his area of concern. For, since he has acquired the scientific mentality he will know how to develop himself and complete his knowledge wherever needed.

The second misunderstanding is seeing the university as an institution of teaching and thus it is believed that the aim of the university is primarily to teach the student contemporary scientific knowledge. This eventually leads to arrange the university not according to an educational philosophy and a theory. As a result no thinker attempts to develop such a theory. We must understand that *theory* is the plan of an achievement. If there is no plan then the whole institution is simply copied from an existing model. This is why most developing nations copy a model of the university from a Western country and try to develop it in their own culture. They cannot develop because the university model in the West is developed on the basis of an educational philosophy which is foreign to the indigenous culture. Since they do not know that philosophy they do not know how to cope with its problems either. I would like to point out that this second problem is mainly dependent on the first one and thus can be solved on the basis of scientific mentality only.

4.2 *External challenges*

There are many external challenges facing higher education today in the Muslim world. There is first the economic challenge because the budget reserved for education is very low in most of these countries. There is a political problem also in the sense that there is hardly any scientific freedom, let alone freedom of thought and expression. Obviously no scientific mentality is possible under these circumstances. But there is one more challenge, which overshadows all these external challenges and that is the moral challenge. Human determination can overcome all economic and political and even social problems with struggle, but the moral challenge can be faced effectively only with the proper moral attitude. Because if its importance I would like to evaluate this challenge by way of concluding remarks and new prospects in higher education.

5 CONCLUDING REMARKS AND NEW PROSPECTS

I consider the moral challenge to be the most outstanding danger facing higher education today. We may deal with other challenges as we can easily recognize them since they are internal to education. But the moral problems facing us are not directly related to education and as a result cannot easily be diagnosed. Therefore, I would like to evaluate the moral challenge as the most conspicuous challenge facing the university today. I would like to illustrate this in the process of the emergence of scientific traditions. For, the moral element is almost totally neglected in analysis of scientific traditions. Therefore, if I evaluate how scientific traditions emerge in history we will be bale to see the function of the moral element. Since the university as an institution represents scientific tradition, then the moral challenge for higher education can also be obtained within the analysis of the emergence of scientific traditions. In this way I am hoping to introduce at the same time scientific attitude embedded in scientific traditions as the new trend in higher education. I emphasize this point because unfortunately it is neglected in so many new universities especially in the Muslim world.

We have already clarified our point that there is a Knowledge Structure in a worldview which directly acts as the ground of all scientific activities. But in this Knowledge Structure there is a need for another conceptual basis to support these activities directly and that has been identified as "Scientific Conceptual Scheme". The emergence of a Scientific Conceptual Scheme within a given society leads to a scientific tradition. Therefore, the very concept of scientific tradition involves the assumption of a scientific community because the rise of a scientific tradition necessarily assumes the existence of a society. In this sense, the idea of scientific tradition which is primarily a cognitive scheme includes the achievements of generations of scientists within which scientific education is carried out and thus supplies a foundation for their further scientific practices. Therefore, this whole process can be analysed in order to reveal its basic factors and phases.

Obviously there must be certain conditions in a society so that science and learning will flourish. However, these conditions cannot so easily be ascertained. That is why we deem it necessary to study them in this context in order to show that there must have been some conditions at the social level, with all its aspects, for the rise of learning and education in a given society. Since these conditions are the causes for the rise of learning within a certain social and cultural context, I would like to call them "contextual causes" for the rise of scientific tradition. It is possible to distinguish certain contextual causes as more rudimentary and hence necessary for the emergence of any kind of scientific activity; we shall refer to such necessary elements leading to the rise of a scientific tradition as 'nucleus contextual causes'. All other peripheral elements that contribute to the rise of the nucleus contextual causes leading to the emergence of science and a scientific tradition can be termed 'marginal contextual causes'.

A nucleus contextual cause is a dynamism which manifests itself at two levels: the first is at the social level, which causes certain unrest and stirring within the society as if the whole structure of the society is re-shaping itself and thus every social institution is affected by this dynamism; but most importantly, the political and educational institutions are re-organized as a result of this unrest; second is at the level of learning (education) and it is this dynamism which causes a lively exchange of ideas on scientific and intellectual subjects among the learned of the community. For instance, in the case of Islam it was internally generated by the thought of the Qur'ân through its dissemination within the first Muslim community. But here what we are trying to look for is whether there is any universal rule (or rules) governing the generation of that dynamism.

At the social level two phenomena may be distinguished as corresponding to the nucleus contextual causes for the emergence of a scientific tradition: the first is moral dynamism; and the second is intellectual dynamism. Therefore, there are primarily two nucleus contextual causes: one is moral in character, the other is intellectual, both of which refer to a dynamism in a given society. With respect to moral dynamism it is possible to divide the members of a given society into three groups:

1. morally sensitive people;
2. the common mass;
3. the selfish or morally insensitive people.

Among these three classes only the moral and the selfish are dynamic. For the former class struggles to restore morality and good order in society, whereas the selfish remain indifferent to this end by spending their dynamism to their own ends. The masses, on the other hand, are driven to either side, which may lead to a struggle on behalf of both sides to defend their goals that may or may not result in intellectual dynamism. This is because the nucleus contextual cause is not the only cause of intellectual dynamism; for this development can be attained only when all other conditions are also present. But if the morally sensitive class becomes victorious and draws the masses towards that end, then intellectual progress can take place once the second phenomenon of the nucleus contextual cause, i.e., intellectual dynamism, is present.

We would like to posit here a feature of scientific attitude, namely originality, which also encompasses dynamism inherently. We may illustrate this point with the Greek scientific tradition. We claim that if there were not in each case a new and fresh outlook, the intellectual dynamism would not have flourished and thus the flair of Greek intellectualism would have died out long before Plato. Moreover, just because there is hardly any original theory and doctrine after Aristotle, Greek intellectualism began to decline after his era. The same is also true for both the Islamic and Western scientific traditions, but the way this intellectual dynamism is manifested in all these societies varies. The 'dynamism inherent within originality and novelty' (of ideas and doctrines) is what we call 'intellectual dynamism'. We consider this also as a natural element in human constitution. What we are showing here is the idea that originality inherently possessing dynamism contributes essentially to the rise of scientific traditions. In fact, originality is invigorating, fascinating and enlivening, it is just like the re-awakening of a land from the demise of winter, and this dynamism is reflected thereby in the society, which is then set into a process of scientific advancement provided that there are no impediments in the way of mutual companionship between science system and its community. Thus without intellectual dynamism no intellectual development is ever possible. But this does not mean that, as we have already stated, once there are original theories and philosophical systems, then such progress will necessarily take place. The reason for this is the other condition of the society, namely, moral dynamism, which must conform to the originality of intellectualism and thus enable it to flourish. Otherwise, intellectual progress will soon die out, which was the case of Greek intellectualism after Aristotle, who is the most original Greek philosopher. But his originality was not sufficient to provide continuity to the apex of Greek thought. In history this has somehow been the ill-fate of all civilizations; a community or nation at the apex of its civilization becomes 'worn out', being burdened by the tremendous weight of its history it begins to decline. Although we do not think that this is a necessary development of a civilization, *viz.* born out-progress-apex-decline and fall, this seems to have been the course of all past civilizations. If our view here concerning the course that intellectual progress and a civilization takes at its rise, is granted then the opposite course will be the natural process of decline, which means that as long as the contextual causes are kept alive the civilization carrying the scientific tradition will continue to live and progress.

One should be reminded that these contextual causes cannot exhaustively be enumerated for all societies. There may be, for example, ten such causes needed in the case of Greek civilization, but this number may be eighteen for another society. Hence, although the number of the nucleus contextual causes as necessary elements may be precise for all societies, the general number of contextual causes, namely, the nucleus and the marginal contextual causes taken together cannot be determined in a decisive manner. In fact we can give examples from the Western scientific tradition that there were thinkers with original ideas and even with novel philosophical systems which did not lead to intellectual progress. Two famous examples are Boethius (d. 524 or 5) and John Scotus Erigena (d. ca. 877). In both of these cases we do not find any continuity of ideas or doctrines after their death, although they both put forward original theories with sufficient vigour. This is because other elements required for intellectual dynamism wcrc not present then and, as a result, we do not see any other scientists furthering their scientific endeavour to lead to a scientific tradition. They will thus remain 'isolated cases' within the history of Western science.

It is possible now for us to elucidate how moral and intellectual dynamism may take place as social phenomena. The moral unrest within a particular society demonstrates a struggle mainly between two classes of people; the morally sensitive and the selfish class. The masses remain as the middle class between the two. When the struggle is taking place, although it is only between the morally sensitive and the selfish, it is immediately passed on to the masses, who become the battle ground of the good and evil forces. Some of the masses are thus won to the moral side, and yet others to the selfish front We may apply here a term from the Islamic civilization, which expresses this social fact: *sunnatullah*. This moral struggle is a *sunnatullah* and thus there is no human society in which this struggle cannot be found in one form or another. When the morally sensitive people exhort sufficient vigour, dynamism and energy, they win to their side an adequate number of the masses and thereby produce intellectual and social dynamism. When the moral struggle between the two groups continues with a victory of the moral class (for this struggle never ends with a victory, but always continues in different forms as long as the society exists), the morally sensitive individuals either produce intellectuals or are themselves intellectuals who formulate original ideas, doctrines and systems by introducing fresh and novel definitions of key concepts that are moral and scientific or knowledge related. This way a lively exchange of ideas and alternative views come into existence within the society; a phenomenon which is necessary to produce intellectual dynamism.

The moral struggle, which is essentially a strife between the good and evil, may either give rise directly to social dynamism, or to intellectual dynamism first, which, then, in turn produces social dynamism. Hence, although in certain cases social dynamism may precede the intellectual one, it does not mean that social dynamism is a nucleus contextual cause for the emergence of a scientific tradition. For the activity in question is of a cognitive nature, namely, science. Therefore, it is still a secondary contextual cause with regard to the nature of the activity in question. But the social dynamism usually leads to an overall activity within the society, which we call 'institutional dynamism'. Hence, there are primarily two marginal dynamisms, which we shall now investigate; *social dynamism* and *institutional dynamism*.

The nucleus contextual causes, i.e., moral and intellectual dynamism, must necessarily produce social dynamism once they are adequately successful. But social dynamism is necessarily preceded by moral dynamism, which we have described quite simply as a moral struggle between the morally sensitive and the selfish; but it is not necessarily preceded by intellectual dynamism. On the other hand, all these various dynamisms are required for intellectual progress that eventually leads to the emergence of a scientific tradition. Yet we distinguish only the moral and intellectual struggles to be the nucleus contextual causes. Since social dynamism is not found at this foundational level, it cannot be included among the nucleus causes. But it must be recognized as a marginal contextual cause. When the nucleus contextual forces are at work, a tremendous social mobility and dynamism begins. It is the dynamism of individuals working together to lead the society as a whole to a morally better situation that we call 'social dynamism', which in turn leads to the re-organization and betterment of social institutions including the political and economic ones as well. It is this reformative and enlightened effort at the organizational level that we call 'institutional dynamism'. When all these contextual causes come together, then they lead the society to intellectual progress. But besides contextual causes different societies may exhibit some other different causes of intellectual progress; such is the case with Western scientific tradition which has Islamic influences also as a cause for the rise of Western intellectualism. Whereas in the Islamic case, the causes are found only within the society, although after the development of Islamic intellectualism in the first century of Islam (i.e., the 7th. century A.D.) it came under foreign influences, especially that of the Greek philosophy and science, which did help improve its intellectualism further.

Since institutional dynamism takes place at the level of social institutions, we must cite them here because of the crucial role they play in the rise of scientific traditions. The most significant of these is the educational institutions; a great reform and re-organization in accordance with the knowledge produced by the intellectual dynamism is required of all the educational institutions, if the society is to produce intellectual progress. Usually there seems to be a relation, although not a necessary one, between the political body and the educational reform. Either the political body brings about the educational reform at the request and direction of the intellectuals, or intellectuals themselves take the initiative and produce educational dynamism, which may in turn lead to a re-organization of the political body and thus produce a great political mobility within the political institutions. These activities which also include the legal undertakings can be called 'political dynamism'. Among these institutional dynamisms we must mention also economic activities. Similar reformations take place in the economic institutions yielding thereby to improve the prosperity of that society and this activity can be called 'economic dynamism'. All these institutional dynamisms do not necessarily develop together within the same period of time and thus helping each other become dynamic reciprocally; or following a different pattern of sequence in every civilization. The educational, political, legal and economic dynamisms include within themselves, with a varying degree of intensity, all the nucleus contextual dynamisms explained above, and as such they are the ones that produce culture. A scientific tradition is born in such a culture only.

We may therefore reduce the factors leading to the rise of a scientific tradition to the following seven factors; moral, intellectual, educational, social, legal and economic. The first two are necessary elements and the last three depend on the educational system. Since higher education represents the scientific attitude reflected in a scientific tradition then obviously the university ought to assume this task. If we see that all these factors are reduced to the moral struggle (ethical *jihâd*) then the scientists in the higher educational institutions must realize where to start. It is for this reason that I am positing the scientific attitude as the most indispensable trend for universities in the Muslim world today.

Higher Education in the Twenty-First Century: Issues and Challenges – Al-Hawaj, Elali & Twizell (eds)
© 2008 Taylor & Francis Group, London, ISBN 978-0-415-48000-0

Enhancement of quality assurance of higher education in Arab countries

Faisal H. Al-Mulla
College of Education, University of Bahrain

ABSTRACT: This paper describes the main feature of review processes and outcomes of the UNDP/Regional Bureau for Arab States project "Enhancement of Quality Assurance and Institutional Planning at Arab Universities". The Project has completed three cycles of programme reviews in the fields of Computer Science (in 15 universities), Business Administration (in 16 universities), and Education (in 24 universities) over a five-year period from 2002 to 2007.

The paper begins with a brief background of the project. This is followed by a detailed description of the basic principles and processes of the Quality Assurance Agency (QAA) academic subject review. All three stages of the QAA review processes (the preparing for the review, the conducting of the review, and the reporting of the review) are defined and outlined.

The paper then considers the main review outcomes (judgments and indicators) of the project in field of education in Arab universities. The paper also considers the common regional issues that are shared by many education colleges across the region.

The paper concludes by outlining some priorities of strategies reform that require collaborative approaches between universities, colleges, departments, and ministries in order to enhance the quality assurance of higher education in Arab countries. With strong market pressures, the greatest demand will be on the quality assurance of higher education in Arab countries.

1 INTRODUCTION

Arab countries have rapidly established a great number of universities in recent decades. In 1950, there were no more than ten universities scattered across the region, whereas today, there are more than 220 higher education providers (UNESCO, 2003). At the same time, Arab countries have, as never before, witnessed a remarkable increase in enrolment rates in higher education institutions. This increase has put more demand on quality of Arab higher education and eventually raises questions of quality assurance. Confirming this, the 1998 Beirut Declaration of the Arab Regional Conference on Higher Education stated "Quality assurance of higher education in Arab countries is under considerable strain, due to high rates of population growth and increasing social demand for higher education, which lead countries and institutions to increase student enrolment, often without adequate allocated financial resources" (UNESCO, 1998, p.44).

Ensuring the quality of higher education is no doubt a major factor for the success of such a mode of education. Accordingly, there is a global trend towards ensuring quality to protect the customer, in this case, the student. In this regard, Alsunbul (2002) indicated that we should go beyond discussing the importance and justification of higher education and focus more on improving its quality, which lies in the totality of all activities that take place during the learning experience. The same attitude was echoed by UNESCO, confirming the importance of ensuring the quality of such programmes to achieve their objectives (UNESCO, 1998).

Quality assurance is in part an administrative review procedure, a series of steps that many universities take in order to review and improve their academic programmes. However, it also has a major symbolic importance, serving as an illustration of the way that academics govern themselves. In the traditional version of programme review, academics take responsibility for defining issues, gathering evidence, and coming to a judgment about the strengths and weaknesses of an academic programme or department. This process exemplifies many values important to academics: encouraging responsible participation of the professoriate in university governance processes; determining matters of educational quality at the departmental level—the "basic unit" of the academic hierarchy; and, most importantly, recognizing the primacy of academic authority on educational matters (QAA, 2004).

Over time, these processes have gained wide credibility, both within the academic and scholarly community and with external regulatory and funding bodies. In the United Kingdom and the United States, most universities and colleges make use of programme review, with most public universities implementing a five-year rolling cycle for review of all academic departments. Similarly, professional licencing bodies in many countries now rely on programme review as an element in their recognition process for educational programmes (Brown, 2004).

Several recent developments suggest that, in many countries, programme review is being expected to take on a new role—as a key component of the quality assessments carried out by governmental and other external agencies. A recent estimate is that there are currently more than 70 quality assessment agencies around the world. In response, new forms of university scrutiny appeared during the 1980s and 1990s in the USA, the UK, continental Europe, and other countries, reflecting concerns about accountability to the public, adequate mechanisms of "quality assurance," or "value for money" (Alderman & Brown, 2005).

The concept of quality assurance in higher education has been around for many years, however it is relatively new to the Arab region where it is just 5 years old (Burden-Leahy, 2005). The United Nations Development Programme (UNDP)/Regional Bureau for Arab States project (RBAS), "Enhancement of Quality Assurance and Institutional Planning at Arab Universities" plays a vital role in introducing the quality assurance mechanisms in Arab region higher education. It seeks to assist a core group of leading public and private Arab universities to adopt internationally-based instruments of quality assurance enabling universities to evaluate programmes, evaluate student performance; and establish statistical databases of participating universities to provide region-wide indicators of programmes, staff, student demographics and university resources and activities (UNDP/RBAS, 2005).

A reading of the three cycles of project reviews commissioned by the UNDP (programmes of Computer Science, Business Administration, and Education) reveals how quality assurance is gaining its form and how the concept is understood in Arab higher education. A brief look at how the UNDP project has started and how principles and processes of Quality Assurance Agency (QAA) academic subject review have been implemented will provide a background for understanding the particular context for implementing quality assurance mechanisms in other Arab countries.

The main purpose of this paper is to describe the main feature of review processes and outcomes of the UNDP/RBAS project, "Enhancement of Quality Assurance and Institutional Planning at Arab Universities". To fulfil this main purpose the following sup-purposes were addressed:

- To provide an overview of the UNDP/RBAS project;
- To identify the main feature of QAA academic subject review;
- To define the main stages of the UNDP/RBAS/QAA academic subject review;
- To present main outcomes of the UNDP project in field of education.

UNDP/RBAS through its Higher Education Project for the Enhancement of Quality Assurance and Institutional Planning in Arab Universities is the only international organization actively engaged in promoting international instruments of quality assurance at the regional level through the evaluation of programmes, the assessment of student performance, and the establishment of comparable statistical databases of participating universities. The Project's development objective is the introduction of independent systems of quality assessment of programmes in Arab universities, with reference to internationally accepted criteria, procedures and benchmarks (UNDP/RBAS, 2006).

In the Arab Region, UNDP/RBAS is currently the only international organization actively engaged in institutional planning at the regional level, expending over USD 4.5 million (BHD 1.69 million) in quality assurance activities. Since its inception in 2002, the Project has reviewed 54 university programmes in 13 Arab countries. The programme of reviews in 2005–2006 addressed selected academic programmes in the field of Education and follows two successful cycles of reviews organized by the Project in partnership with Arab universities: Computer Science programmes in 2002–2003 and Business Administration programs in 2003–2004.

The method used for review is a modified version of the Academic Subject Review as developed for implementation in 2000 by the Quality Assurance Agency for Higher Education (the QAA) in the United Kingdom. This method is itself a direct development of the earlier Subject Review method used to review academic disciplines at UK universities over the period 1992 to 2001.

Academic subject review takes place according to the Project's Handbook. It places responsibility on the university to evaluate and report on the academic standards of its programmes of study and the quality of learning opportunities. This evaluation takes place within the agreed framework for review. This framework, described in the Project's Handbook, includes the use of external reference points to establish and improve the academic standards.

With regard to the reviews carried out in the context of this Project, each university was asked to identify its external subject reference points so that its academic standing could be judged. Participating universities were also invited to determine the extent of overlap between their curricula and the published topics of the Major Field Test in Education (MFT), which is developed and managed worldwide by the Educational Testing Service (ETS) based in the USA. To this end, each university carried a detailed cross-matching analysis between the topics of the MFT and the curriculum of the reviewed programme (UNDP/RBAS, 2006).

The Project has completed three cycles of programme reviews in the fields of Computer Science (in 15 universities), Business Administration (in 16 universities), and Education (in 24 universities) from 13 Arab countries, over a five year period from 2002 to 2007. The Project's fourth cycle of reviews, targeting programmes of Engineering, is planned for 2007–2008. The size of this cohort is expected to exceed 100 universities, distributed over 14 Arab countries. The project has produced the first Arab cohort of experienced and qualified quality assurance reviewers, training over 71 representatives in all stages of the review process, of whom 62 participated in external review missions to other universities in the first three cycles.

Currently and with combined Finnish and German partnership and co-sponsorship, UNDP/RBAS is embarking on institutionalizing the work and objectives of the regional project in the form of an Arab Regional Quality Assurance Agency (ARQAA). ARQAA's mandate will be to assist the region's universities, through active regional platforms, to develop and implement common methods and standards of quality assurance.

2 FEATURES OF QAA ACADEMIC SUBJECT REVIEW

Quality assurance refers to a set of approaches and procedures regarding the measurement, monitoring, guaranteeing, maintenance, or enhancement of qualify of higher education institutions/providers and programmes, or processes by which the achievement of education programme standards, established by institutions, professional organizations, government and other standard setting bodies, is measured (Lemaitre, 2004).

The feature of QAA subject review is a peer review process. It starts when institutions evaluate their provision in a subject in a self-evaluation document and prepare a programme specification for the selected award-bearing programme. The self-evaluation and the programme specification are submitted to the Agency for use by a team of reviewers. The reviewers are academics and are practitioners trained and experienced in external scrutiny and review processes. They read the documents and visit the university to gather evidence to enable them to report their judgments on the academic standards, the quality of learning opportunities and the ability of the university to assure and enhance academic standards and quality. Review activities include meeting staff and students, scrutinizing students' assessed work, reading relevant documents, and examining learning resources. The team gives an oral summary report to the university at the end of the review visit and prepares the written report (QAA, 2004).

The review procedure looks at the effectiveness of an institution's quality assurance structures and mechanisms, and at how the quality of its programmes and the standards of its awards are regularly reviewed, and the resulting recommendation implemented. It also looks at accuracy, completeness and reliability of the information that an institution publishes about quality and standards (Goodison & Lewis, 2005).

2.1 *The stages of the UNDP/RBAS/QAA academic subject review process*

The review process comprises three stages (QAA, 2004):

- preparing for the review;
- conducting of the review;
- reporting of the review.

2.1.1 *Stage 1: Preparing for the review*

The process begins with the Scope and Preference Survey with institutions supplying the UNDP/RBAS Project Manager with information about their programmes. This information is used by the Project Manager to establish a program of reviews and to select the proposed review teams. On completion of any further enquiries, the UNDP/RBAS Project Manager opens a dialogue with each institution, involving such matters as the appropriate timing of the review.

The institution submits its self-evaluation document (SED), including programme specifications, to the UNDP/RBAS Project Manager no later than two months before the agreed initial meeting date. The self-evaluation should be submitted to the UNDP/RBAS Project Manager with a standard cover sheet. Once the UNDP/RBAS Project Manager has checked the cover sheet, the SED is sent to the Review Co-ordinator (the review team leader, usually from UK) who analyses it against a standard template to ensure that it forms an appropriate basis for the review to proceed. The UNDP/RBAS Project Manager, the officer with responsibility for the proposed review, will check the self-evaluation to ensure that it has the required contents and that a review can proceed. When the UNDP/RBAS Project Manager has notified the institution that it accepts the SED as the basis for review, the institution sends copies of the SED to the review team members (reviewers include both UK and Arab reviewers). The specialist reviewers read and comment upon the SED; the Review Co-ordinator uses their comments to help plan and set priorities for the review.

The SED is central to the process of subject review, and fulfils two main functions: it is intended to encourage subject providers themselves to evaluate the standards achieved by students and the quality of the learning opportunities offered to them; and it provides a framework for a process of subject review based on the testing and verification of statements made by subject providers. It is a statement which demonstrates that a subject provider has evaluated, in a constructively self-critical manner, the following issues:

- the appropriateness of the academic standards it has set for its programmes;
- the quality of the learning opportunities provided for students;
- the effectiveness of the mechanisms and procedures in place for assuring and enhancing the standards and quality of the provision.

The SED should discuss both strengths and weaknesses of provision, as perceived by the provider. The document allows the provider to demonstrate how the strengths of the provision identified in previous internal or external reviews or accreditation events, where applicable, have been built upon, and how any weaknesses identified have been addressed. Where weaknesses remain, plans for addressing these should be summarized. The SED supports by two documents: the Programme Specification(s) and the Appendix of Key Information and Data.

2.1.2 *Stage 2: Conducting the review*

The main period of review activity normally lasts eight weeks from the institution submitting its SED. Reviewers will spend some of this time in the institution. It is essential that reviewers are able to gather sufficient evidence to allow them to test statements made in the SED, and to form robust judgments on the quality and standards of the provision. Reviewers reach collective judgments at the judgment meeting at the end of this period.

A number of key meetings are held during the review period. These will be with subject and other staff from the institution, current and former students, and, where appropriate, employers. The subject review facilitator, appointed by the institution, will be invited to attend all reviewers' meetings except those with current and former students, employers and meetings where judgments are discussed.

First review team meeting: The reviewers, together with the subject review facilitator, will meet before the initial meeting to discuss the review schedule and to share their early perceptions. The Review Co-ordinator will ensure that there is a shared understanding of the nature and purpose of the review. The reviewers will agree key questions for discussion with staff.

Initial meeting with the institution: With the agreement of the reviewers, the subject provider may make a brief presentation, typically no more than 10 minutes, to introduce the provision to be reviewed and to describe any developments since the SED was prepared. The Review Co-ordinator will remind both the reviewers and the institution representatives of the method and protocols of review and the schedule agreed so far.

Other meetings: Other meetings will be arranged with staff to discuss academic standards, the quality of learning opportunities and the quality assurance and enhancement. There is no fixed pattern of meetings, as the reviewers and the institution will need to agree a plan for each review which enables the reviewers to gain the evidence they need to make judgments with minimal disruption to the institution. The institution may wish to consider who is appropriate to attend these meetings. The review may also include meetings with current students, former students, employers and work placement providers.

Testing the self-evaluation and gathering evidence: The reviewers have a collective responsibility for gathering, verifying and sharing evidence so that they are able to test statements made in the SED and develop judgments on quality and standards.

Documents are important sources of evidence that assist the reviewers to evaluate the quality of learning opportunities and academic standards achieved. Documentary evidence includes student work, internal reports from committees, boards and individual staff with relevant responsibilities; and external reports from examiners, verifiers, employers, validating and accrediting bodies. Reviewers also gain evidence from observing some elements directly to evaluate their quality, for example, learning resources.

The reviewers are expected to identify, share, consider and evaluate evidence related to the programmes under review. The reviewers keep notes of all meetings with staff and students, of their observations, and of comments on the quality of student work and its assessment. These should be analytical rather than descriptive, and refer to sources of information as well as to direct observations. Strengths and area for improvement are summarized. Circulation of notes between the reviewers, and collation of notes by the Review Co-ordinator, will assist the reviewers to develop a collective evidence base on which judgments can be made.

Making judgments: The reviewers will arrange to meet in order to arrive at their final judgments; this is usually in the last day of their visit to the institution. The reviewers will share and consider all forms of evidence gained during the review to enable them to arrive at accurate and robust collective judgments. Judgments are usually made in the three review areas:

- academic standards (intended learning outcomes, curriculum, assessment, and student achievement);
- quality of learning opportunities (teaching and learning, student progression, and learning resources); and
- quality assurance and enhancement.

Judgments about academic standards are made on the appropriateness of the intended learning outcomes set by the subject provider in relation to subject benchmark statements, qualification levels and the overall aims of the provision; on the effectiveness of curricular content and assessment arrangements in relation to the intended learning outcomes; and on actual student achievement.

Judgments about the quality of learning opportunities are made on the effectiveness of teaching and the learning opportunities; on the effectiveness of learning resources, including staff; and of the academic support provided to students to enable them to progress in their studies.

Judgments about the quality assurance and enhancement are made on the arrangements for assuring and enhancing the provision quality.

Each standard of the three review areas will be judged on one of three judgment-criteria: **Good, Satisfactory,** or **Unsatisfactory**. A **good** judgement will be made if the strengths of the provision substantially outweigh any minor weaknesses. A **satisfactory** judgement will be made if where the provision passes the minimum threshold but nevertheless has some significant weaknesses. An **unsatisfactory** judgment will be made if the institution is unable to provide sufficient evidence for the reviewers to reach a "satisfactory" or "good" judgment.

Judgments on academic standards: Reviewers make a single, threshold judgment about academic standards. They take into account the points set in the Review Handbook to decide whether they are satisfactory or not with the academic standards of the provision under review. Each of the four standards will be evaluated, but no formal judgment will be made on Intended Learning Outcomes. If any of the other three standards is judged unsatisfactory, academic standards overall will be judged *unsatisfactory* (Goodison & Lewis, 2005).

The reviewers will assess, for each programme, whether there are clear intended learning outcomes that appropriately reflect a range of reference points: subject benchmark statements and the level of the award. The reference points are provided to assist reviewers in determining whether provision is meeting the standards expected by the academic community generally, for awards of a particular type and level. If the reviewers find that the intended learning outcomes do not match those expectations, it is unlikely that they are satisfied with the standards of the provision.

The reviewers will assess whether the content and design of the curriculum are effective in enabling students to achieve the intended learning outcomes for the programme. Institutions should be able to demonstrate how each outcome is supported by the curriculum. The reviewers will assess whether the curriculum content is appropriate to each stage of the programme, and to the level of the award. Institutions should be able to demonstrate how the design of the curriculum secures academic and intellectual progression by imposing increasing demands on the learner, over time, in terms of the acquisition of knowledge and skills, the capacity for conceptualization, and increasing autonomy in learning. If students cannot develop significant intended learning outcomes through the curriculum, it would be unlikely that the reviewers are satisfied with the standards of the provision (QAA, 2004).

The reviewers will evaluate whether assessment is designed appropriately to measure student achievement of the intended learning outcomes. Institutions should be able to demonstrate how student achievement of intended learning outcomes is assessed, and that, in each case, the assessment method selected is appropriate to the nature of the intended learning outcomes. Satisfaction with the security and integrity of the assessment process, with appropriate involvement of external examiners, is essential. The range of assessments planned should include some that have a formative function and provide students with prompt feedback to help them to progress in their studies, and assist them in the development of their intellectual skills. There should be clear and appropriate criteria for different classes or levels of performance, and these criteria should be communicated effectively to students. If significant intended learning outcomes are not assessed, or if the reviewers have serious doubts about the integrity of the assessment procedures, it would be unlikely that they are satisfied with the standards of the provision (QAA, 2004).

The reviewers will assess whether student achievement matches the intended learning outcomes and level of the award. The reviewers will consider external examiners'/verifiers' reports from the three years prior to the review, and will also sample student work. The balance between reliance upon the reports of external examiners'/verifiers' and direct sampling of student work will depend on the confidence that the reviewers have in the internal examining and verification arrangements of the institution. Review reports will include comments on strengths and areas for improvement for each standard of the academic standards judgment.

Judgments on quality of learning opportunities: Reviewers will make judgments about the quality of the learning opportunities offered to students against the broad aims of the provision and the intended learning outcomes of the programmes. Each judgment will normally cover all provision within the scope of the review. However, if performance is significantly different in a subject area, or for a particular award, separate judgments will be made.

The reviewers will rely usually on secondary evidence rather than direct observation of teaching and learning, wherever possible. The reviewers will assess the effectiveness of teaching and learning in relation to curriculum content and programme aims.

The reviewers will evaluate student progression by considering their recruitment, academic support and progression within the programme. In making judgments about learning resources, the reviewers will assess whether the minimum resource necessary to deliver each programme is available, and will then consider how effectively resources are used in support of the intended learning outcomes of the programmes under review. The reviewers will look for a strategic approach to the linkage of resources to intended learning outcomes for each programme.

Quality assurance and enhancement: Institution-wide systems for the maintenance and enhancement of standards and quality are addressed through the institution's arrangements for programme review and evaluations. The reviewers will gather evidence, not least from their discussions with staff and students and their scrutiny of external examiners' reports, on the operation of the institution's systems in each subject under review. The final section of the review report will express their evaluation of the ability of the institution to maintain and enhance standards and quality in the particular subject (Goodison & Lewis, 2005).

2.1.3 *Stage 3: Reporting of the review*

At the end of the review period, the Review Co-ordinator will usually notify the institution representatives and staff orally of the judgments reached. This takes place normally in the last meeting. At the end of review period the Review Co-ordinator will prepare a draft report (approximately 4,000 words) drawing upon the contributions of the specialist reviewers. The draft report will be edited before it is sent to the institution through the UNDP/RBAS Project Manager, normally within two months of the end of the review period. The institution is asked to comment upon matters of factual accuracy and return these to the UNDP/RBAS Project Manager within three weeks of receipt. Following the receipt of these comments, a further draft is prepared by the RBAS Regional Director and forwarded to the presidents and programme providers of the participating institutions. The report will usually include judgments reached as well as common issues revealed by the assessment process of the provision including patterns of strength and weakness, as well as lines of needed reform.

2.2 *Outcomes of the UNDP/BRAS project in the field of education*

Overall, the outcomes of the project in the field of education in Arab universities find that the participating universities are making a strong contribution to the training of graduates and postgraduates in Education, who meet the specific requirements of school-based education, in the various Arab states and generally are well regarded by the schools and other employing bodies, including the various Ministries of Education.

Academic standards were judged to be good in five of the participating universities, satisfactory in 16 and unsatisfactory in two. Wide-ranging and relevant curricula were generally a point of strength, as was student achievement at both the undergraduate and postgraduate levels. However, assessment continues to be a weakness, particularly because of its emphasis on memory recall of descriptive knowledge, its lack of focus on higher-level cognitive skills and the absence of internal or external mechanisms to ensure its fairness and transparency. Institutional systems for student support are not strong and targeted support for students with various special needs is patchy.

The quality of teaching and learning opportunities also rates as generally strong—judged to be good in 16 universities and satisfactory in seven. Despite an over-reliance on set lectures, the assessors noted a wide range of innovative teaching methods. Learning resources (including space accommodation, libraries, ICT, and media facilities) were judged to be good in 11 universities and inadequate in the others especially with respect to ICT equipment, the number of personal computers available for student use, and access to inter- and intra-net facilities.

Quality assurance and enhancement continue to be the weakest aspects. According to the outcomes of the project, six universities were judged to be unsatisfactory as opposed to 13 satisfactory and only four judged as good. These figures, however, represent an improvement since the HE Project's first report, which assessed academic programmes in Computer Sciences. Fully articulated university quality assurance systems are still in the minority, though a serious effort has been initiated in eight universities. A quality culture in which annual feedback, evaluation and monitored action plans are the norm has not yet evolved.

The outcomes of the project in the field of education propose a strategic reform agenda comprising nine priority recommendations including: 1) adoption of a pro-active strategic approach to curriculum design based on Intended Learning Outcomes; 2) allowing greater academic freedom in curricula and reducing admission controls; 3) use of external references and inputs to benchmark effectiveness and ensure compliance with international standards; 4) greater emphasis on higher-level cognitive skills such as evaluation, critical analysis and synthesis; 5) adopting a more pro-active approach to staff training and development for teaching, learning and assessment; 6) improving student support systems with special emphasis on vulnerable students with special needs; 7) upgrading the quality and availability of learning resources, particularly ICT facilities; 8) formalizing and institutionalizing quality assurance and enhancement systems; and 9) improving teaching and learning resources in Arabic and supporting the enhancement of language skills to allow students to take full advantages of foreign resources.

These outcomes reiterate the main conclusion of its two predecessors that a regional initiative to adopt and implement the nine steps of this strategic reform agenda should continue to be a shared priority for Arab policymakers in higher education (UNDP/RBAS, 2006).

3 CONCLUSIONS

The main purpose of this paper has been to describe the main feature of review processes and outcomes of the UNDP/RBAS project, "Enhancement of Quality Assurance and Institutional Planning at Arab Universities".

The Arab region has witnessed a remarkable increase in the number of higher education institutions over the past two decades. This significant progress, however, does not necessarily denote quality programmes. A key problem is how to ensure that a quality learning experience is being provided. Thus, the need for developing quality assurance frameworks for higher education institutions and programmes in the Arab region is evident.

It appears that the UNDP/RBAS is the only international organization actively engaged in promoting international instruments of quality assurance at the Arab Region through the evaluation of programmes, the assessment of student performance, and the establishment of comparable statistical databases of Arab universities. The Project provided the participating universities with an opportunity to undertake a comprehensive review of their Education programmes, and receive a site visit from external reviewers together with an oral feedback report, and at a later stage, a written review report.

The evidence suggests that the quality assurance and enhancement continue to be the weakest aspect of provision in Arab universities. Most universities in Arab countries do not display any mechanisms of quality assurance in the reviewed programmes. This means that quality assurance activity is non-existent in many universities. Therefore, there is a need for Arab universities to place more explicit emphasis on the importance of the quality assurance mechanism.

Finally, given all the issues highlighted in this paper, it seems honest to conclude that adopting or developing such a framework is essential for ensuring quality of higher education in Arab countries, but is not enough on its own to ensure quality higher education offerings. It should be regarded only as the beginning of an on-going comprehensive process that has to be supported effectively to produce the desired results. In conclusion, the stronger market pressures, the greater demand will be on the quality assurance of higher education in Arab countries.

REFERENCES

Alderman, G. & Brown, R. 2005. Can quality assurance survive the market? Accreditation and Audit at the crossroads. *Higher Education Quality*, 59(4): 313–328.

Alsunbul, A. 2002. Issues relating to distance education in the Arab world. *Convergence*, 35(1): 59–80.

Brown, R. 2004. *Quality Assurance in Higher Education: The UK Experience Since 1992*. London: Routledge Flamer.

Burden-Leahy, S. 2005. Addressing the tensions in a process-based quality assurance model through the introduction of graduate outcomes. *Quality in Higher Education*, 11(2): 129–135.

El-Khawas, E. 1997. Linking programme review to external scrutiny: International developments. *International Higher Education*, November (9), 13–15.

Goodison, R. & Lewis, D. 2005. *Handbook for Academic Subject Review*. A handbook prepared for the UNDP higher education project UNDP/RBAS/RAB 2001–2002.

Quality Assurance Agency for Higher Education (QAA), 2004. *Handbook for Academic Subject Review: England*. Retrieved 16 April 2007, from www.qaa.ac.uk.

Samoff, J. 2003. Institutionalizing international influence, in: *Comparative Education: the Dialectic of the Global and Local*, Arnove, R. F. & Torres, C. A. (eds.): 52–91, Rowman & Littlefield publishers, New York (2nd edition).

UNDP/RBAS, 2005. *Quality Assessment of Computer Science and Business Administration Education in Arab Universities: A Regional Overview Report*. Retrieved 12 May 2007 from http://rbas.undp.org/site_docs/HE%20Overview%20Report.pdf

UNDP/RBAS, 2006. *Quality Assessment of Programs in the Field of Education in Arab Universities: A regional Overview Report*. Retrieved 15 May 2007 from http://rbas.undp.org/site_docs/UNDP-Final_Printed.pdf

UNESCO, 1998. *Higher Education in the Twenty-first Century: Vision and Action*. World Conference on Higher Education, Final Report, volume 1. Paris: UNESCO.

UNESCO, 2003. *Higher Education in the Arab Region, 1998–2003*. Meeting of higher education partners. A document prepared by the UNESCO Regional Bureau for Education in the Arab States, 23–25 June 2003. Paris: UNESCO.

Higher Education in the Twenty-First Century: Issues and Challenges – Al-Hawaj, Elali & Twizell (eds)
© 2008 Taylor & Francis Group, London, ISBN 978-0-415-48000-0

Building capacity for labour-market flexibility in a globalized world: The role of universities

Amer Al-Roubaie
Ahlia University, Kingdom of Bahrain

1 INTRODUCTION

In recent decades, globalization has given rise to a new economy driven by factor mobility, skills development and knowledge-creation. No longer are capital and labour alone sufficient for promoting global competitiveness; instead, knowledge-creation and development of new products that induce global integration are requisite. The new knowledge-based economy underscores the importance of information and knowledge driven by labour market flexibility and investment in modern Information and Communications Technologies (ICTs). To this end, deepening integration in global markets and cultivating the benefits of globalization will require a highly-skilled and well-trained workforce capable of making strategic, managerial and organizational decisions.

In the new economy, universities have become indispensable for building knowledge capacity that accelerates economic growth and increases global linkages. Managing globalization mandates skill development and on-the-job training generating the human capital necessary to achieve labour supply fluidity. Integration in the global economy also requires the invention of new production methods and the creation of new products. Harnessing of universities by governments to build national capacities entails universities' designing research programs and offering special courses to increase human capital and to satisfy labour demands of globalization. Increasing society's stock of knowledge generates labour flexibility that enhances global integration.

The primary aim of this paper is to examine the relationship between the new economy driven by globalization and the potential contribution of universities to labour market flexibility especially in reference to Arab countries generally and the Gulf Cooperation Council (GCC) states particularly. It has been widely acknowledged that promoting global competitiveness requires greater labour market flexibility in order to ensure managing globalization and increasing global linkages. In Arab countries, however, labour markets are rigid due to government laws, market imperfections, institutional problems and absence of gender empowerment. These rigidities mandate that radical reforms be taken to improve market flexibility and increase global linkages. Efforts to facilitate global integration are being made by GCC countries *via* building technological capacity and investing in human capital.

To gain perspective in the role of universities in building capacity through enhancing labour market flexibility, a prefatory discussion of global interdependence and the knowledge-based economy is warranted.

2 GLOBAL INTERDEPENDENCE

Technological development in a rapidly changing world makes education exceptionally important. Globalization is changing the rules of doing business by shifting attention towards knowledge creation and information dissemination which together induce capacity-building that serves to accelerate economic growth and increase global competitiveness. Globalization offers both opportunities and challenges. However, for most developing countries, neither opportunities nor challenges are within their immediate ambit due to a knowledge deficit, low skill level, inadequate investment, market imperfection and labour immobility. Such conditions are likely to impede global integration and international competitiveness. Global economic trends, currently driven by globalization, are controlled by a few players that exercise substantial leverage over world trade, finance, technology, knowledge and information. The rest of the world, representing mainly non-industrialized countries, is at the receiving end of globalization with little or no say in global decision makings. Recent experience with globalization has shown a wide divergence in both income and knowledge among and within nations and regions. For example, income inequalities between rich and poor countries have widened to the extent that high income countries representing 15 percent of world population earned 80 percent of total world's GDP in 2004. On the other hand, low income countries representing 37 percent of world population earned 3 percent of total world's income in the same year. Such inequalities, in which lack of knowledge perpetuates poverty and deprivation, feed acute instabilities in the world body-politic causing, in the process, injustice, insecurity and violence. Recent United Nations reports on the status of human development and the Millennium Development Goals are important indicators of the complex nature of relations between the South and the North with respect to production, distribution, communication, information, trade, technology and knowledge.

The globalization process is described by increasing interdependencies and interconnection among individuals, groups and nations. However, the multidimensional facets of globalization, has given the term a special importance in the literature on social, political, economic, cultural, international, technological, scientific and educational studies. As a consequence, substantial research, both theoretical and empirical, has been produced about the impact of the new economy on international relations, economic transactions, factor movements, poverty eradication and global understanding. Globalization is also viewed as having political fallout by weakening the sovereignty of the nation-state and reducing its management over the economy. In addition, liberalization and privatization to accommodate globalization may not necessarily, at least in the early stages of development, be feasible to induce integration of the national economy into the global markets. Recent financial and economic crises in Asia, Argentina, Mexico and Russia underline the perils of an overly rapid accommodation to the dictates of globalization where economies were wracked by high unemployment, currency devaluation, balances of payments disequilibrium, asset crashes and private sector bankruptcies. A high degree of global dependency could increase economic instability by subjecting non-industrialized economies to wide fluctuations. Thus, minimizing financial risk and strengthening economic stability will require managing globalization by introducing measures that provide managerial and organizational enforcement and supervision over global activities. Exercising influence over global economic trends mandates greater market access to lessen the downside risks associated with cross border transactions on the national economy. Both opportunity and risk are inherent in the new economy driven by globalization. In this regard, building capacity becomes necessary for increasing flexibility of the economic structure to absorb knowledge and disseminate information. Unfortunately, there are very few countries that can meet such requirements because of the complex nature of global competitiveness and of inefficiencies of existing institutions.

Access to global markets requires radical reforms designed to strengthen national capacity to absorb and create knowledge. In most developing countries, the prospect for building such capacity is constrained by the rigidity of the economic structure, in which the inflexibility of the labour markets features prominently. Although information and communications technologies (ICTs) are facilitating access to global knowledge, the low technological absorption capacity in these countries often relegates their participation in the new economy to by-stander status. Building capacity for knowledge readiness goes far beyond the nation's financial and technological capabilities to include social, cultural, political, institutional and international reforms. Globalization is inducing changes that require reengineering the production structure by increasing (1) the share of high-value- added manufacturing production and exports in relation respectively both to total manufacturing output and exports and (2) the share of the service sector in overall GDP.

As a consequence of these induced changes, however, globalization could create imbalances within the economies of developing countries through polarization and marginalization of various groups and regions within and among nations. For example, India's globalization, characterized primarily by its emergence as an outsourcing and software development centre, has engendered the creation of a highly skilled technocratic class, knowledge-rich in ICTs, in several urban centers in India while bypassing, in the main, poverty in rural India. Such uneven development reflects bias in the distribution of income created by globalization in favor of class linked to global production system. The expertise of this technocratic class transcends "traditional software operations (basic software, data transcription, telephone call centers) to increasingly sophisticated business process outsourcing (BPO).[1] N. R Narayana Murthyy of Infosys underscores that only if ICT development is complemented by "the development of strong infrastructure ... and effective investment in social and educational improvement ... will India be able to integrate ICT in a truly sustainable manner and meet overall development goals" designed to eradicate poverty. (Executive Summary of the Global Information Technology Report 2005–2006, p. XI.)

An important manifestation of globalization is the universal application of ICTs to development, education, knowledge, skills communications, and lifelong learning. This has given rise to the global village. In the global village, except for those excluded by the digital divide, access to knowledge, information, news, cultural products, films, ideas, nationalism and ideologies is within the reach of all people despite their geographical locations and time zones. The new global society is described as the death-knell of geography in terms of its compressing time and space consonant with the rapid interconnections occurring among peoples worldwide. In addition, factor mobility involving capital and, to a lesser extent, labour, are enjoying freer cross-border movement to enable individuals, corporations and nations to reap greater benefits from the new global economy. Today we are witnessing computers "talking" to each other, satellite transmission of data, mobile telephony, voice-of-the-internet protocol (VOIP) communications, universal digital libraries, as well as high speed internet uploads and downloads of information providing instantaneous access to newspapers, films, music and books. Similarly, access to universities and libraries can be had with a user name and password and a click of a mouse. Internet connection could enhance university education by providing face-to-face lectures, virtual laboratories, teleconferencing and electronic examination. The use of various forms of multimedia communications, live videos and bandwidth data networks could provide an effective means for students, lecturers and educators that enhance the quality of education and increase the learning capability of students. In countries where educational institutions are still inadequate to meet the socio-economic objectives, the use of ICTs could have a positive impact on knowledge creation, technology transfer, dissemination of information and sustainable development.

Table 1 illustrates global connections reflecting the share of various regions in global trade and in connectivity through communications. The table shows that among the regions of the world, the Middle East and North Africa (MENA) lags behind in several indicators of globalization. MENA finishes dead last in the key indicator of high-tech exports as a percentage of manufactured exports and, in terms of indicators of global connectivity, ranks superior only to South Asia and Subsaharan Africa. Currently, the share of high-tech exports as a percentage of manufactured exports from Arab countries accounted only for 3 percents. This low percentage of high-tech exports by Arab countries reflects the inadequacy of knowledge readiness in these countries to exploit the opportunities offered by globalization. As in the case of Sub-Saharan Africa, the indicator trade as a percentage of GDP is inflated by the contribution of primary (raw material/energy) exports indicative of the lock-grip of the old economy on these regions. Such tendencies reflect the inability of MENA to build technological capacity that meets the challenges of globalization. In other words, "the region's economic importance and international competitiveness has declined and by failing to reform or operate for change, the countries of the region have failed to reap the benefits of globalization."[2]

---

[1] Jeffrey Sachs, The End of Poverty, (London: Penguin, 2005), p. 182.

[2] United Nations, Economic and Social Commission for Western Asia, Globalization and Labour Markets in the ESCWA Region (New York: United Nations, 2001), p. 22.

Table 1. Globalization indicators, 2004.

Region	Share of exports in total world exports %	High-tech exports (% of manufactured exports)	% Share of service exports in total service export	Internet users per 1000 people	Per 1000 personal computers	Trade (X+M) as % of GDP
East Asia & Pacific	10.6	34	5.8	74	38	71.1
Europe & Central Asia	6.7	9	5.5	138	110	70.9
Latin America & Caribbean	5.1	13	2.8	115	92	44.6
Middle East & North Africa	1.9	3	2.8	42	49	55.1
South Asia	1.1	4	2.0	26	12	27.9
Sub-Saharan Africa	1.6	4	1.1	19	15	54.7
High income	73.0	19	80.0	545	574	41.5
Low & Med. Income	27.0	4	20.0	24	11	37.8
World	100	20	100.0	84	130	44.9

Source: United Nations, Human Development Report 2006 (New York: United Nations, 2006); World Bank, World Development Indicators (Washington: World Bank, 2006).

Highly integrated world markets have given greater power to a few key players, mainly the industrialized countries, transnational corporations and special interest groups, to exercise considerable influence over global trade, finance, technology, and labour flows. For example, the sales of the top 500 multinational corporations (MNCs) known commonly as "The Fortune 500" accounted for about $US13 trillion or close to one third of the global GDP in 2004. The revenues of The Fortune 500 exceed the gross domestic products of all non-industrialized countries *combined*. The new economy is creating a chasm between the initiators of globalization, the *globalizers*, and the *globalized*. The latter group includes mainly developing countries which are 'information-poor and knowledge-poor consumers of the products of globalization.' These downtrodden countries are poorly equipped with the resources, both human capital and technical infrastructure, that would qualify them to profit from globalization.

3 THE KNOWLEDGE-BASED ECONOMY

Rapid connection in the new society requires re-engineering of the communications sector to increase knowledge acquisition. In terms of its jumpstarting human development and accelerating economic growth, investment in education and building infrastructure are considered the *sine qua non* of the new economy. A high quality workforce could lead to a paradigm shift by allowing the country to leapfrog from a low to a high stage of economic development. A key element in this national strategy is to make use of ICTs where possibilities for knowledge production in which return to scale are limitless: knowledge production manifests increasing returns to scale rather than the phenomenon of diminishing marginal returns (which almost all resources exhibit). Knowledge production is a sustainable motor that powers the future development of an economy. In the new economy, knowledge both creates and enables the adoption and adaptation of technologies: individual technologies exhibit a markovian birth-death process but knowledge provides an infinite well-spring generating the birth of new superior replacement technologies where superiority reflects both economic and non-economic variables. Given its vital role in the new economy, knowledge generates wealth creation and poverty eradication. The digital divide between poor and rich nations is measured by the knowledge gap and global connectivity. A nation's drive to achieve comparative advantage is no longer measured by the size of its factor endowments or financial capabilities; rather, by its capacity to apply, absorb and create knowledge.

The new economy, driven by globalization, requires knowledge workers trained in IT. However, lack of adequate knowledge of ICTs could become barriers to overcoming the challenges of the global economy by preventing a given country from applying and absorbing global knowledge and information. Production for global markets requires a variety of information about production, distribution, consumption, investment, finance and transnational business. Access to information underscores the importance of building technological capacity to enable linking the local economy to global markets. Local manufacturers and producers need to adopt new technologies if they wish to gain from the new opportunities afforded by globalization. At least as a short-run expedient, technologies often compel firms to depend on external labour to meet their labour supply requirements in the new economy. Inculcating new skills in the domestic labour force will depend on the adequacy of infrastructure as well as on the capacity of the economy to absorb global knowledge and technologies. In terms of job security, only a small number of the new jobs may be classified as full-time while the outsourcing-to-"in sourcing" ratio may be adverse even more so where an economy is more a recipient of outsourcing services rather than a deliverer. Such trends could create problems for many of the developing countries which suffer from a lack of a skilled workforce especially when combined with inadequacy of financial resources and inefficient institutions. Such inefficiencies hobble global access to knowledge and information.

During the last several decades, the source of economic growth in most industrialized countries has been attributed to investment in science and technology. Economic transformation requires new techniques and new products in order to stimulate productivity and growth. Rapid transformation of the Asian economies in the 1970s and 1980s has been linked to the increase in total factor productivity rendered by heavy investment in human capital and global access through ICT.

A substantial portion of goods and services traded in the global markets contain a high content of skills and knowledge; succinctly expressed, these goods are skill- and knowledge intensive. *Ipso facto* the very production of these goods implies that globalization is increasing demand for labour with specific qualifications including web designers and masters, wireless developers, knowledge engineers, technology security specialists and data miners. The use of highly sophisticated computer technologies in industry and services has increased the demand for specific skills needed to enhance productivity. In other words, globalization is increasing specialization in production which mandates creation of more flexible productive structures in order to meet market changes driven by globalization. Under such circumstances, the economic success of a country will depend on the ability to build capacity to disseminate information as well as on the absorptive capacity to acquire, apply and create knowledge. In the new economy, comparative advantage depends on labour quality, knowledge management, organizational skills and the capacity to acquire and create knowledge. Overcoming such challenges will require the developing countries to adopt a "selective approach that aims at striking a careful balance between opening and protecting the domestic market [rather than blindly opting for] rapid liberalization"[3] without due care for the deleterious side-effects generated by the latter path. Globalization must be managed effectively if developing countries are to benefit from it. The process of liberalization should not be endorsed without careful assessment of the ability of the economy to adjust to the new changes. The country must be capable of developing a technological capacity that allows greater flexibility in terms of knowledge absorption and production of new products.

---

[3]United Nations, Ibid, p. 19.

4 BUILDING TECHNOLOGICAL CAPACITY

Building technological capacity is a complex process involving governments, universities, private enterprises, multinational businesses, technology transfers, international institutions and non-governmental organizations. The role of the government transcends budgetary expenditures and incentive programs to include: contributing directly to R&D, building infrastructure to attract FDI, encouraging technology transfer, establishing institutions and supporting local enterprises against monopolies and market failures. The knowledge-based economy manifests production of new techniques, new products and new knowledge to enable the country to be to compete viably in diverse global markets. Building sound infrastructure, establishing effective institutions, providing adequate funding, and promoting freedom represent important indicators of knowledge-readiness and technological advancement. Such capacity enhances an economy's ability to innovation and to develop new products. As pointed out by the United Nations: "Knowledge acquisition entails not only building on a country's own knowledge base to generate new knowledge through R&D but also harnessing and adapting knowledge available elsewhere through openness, broadly defined, including, e.g., promoting the free flow of information and ideas, establishing constructive engagement in world markets, and attracting foreign investment."[4]

Technological innovation engenders a positive impact on labour productivity by employing more ICT through altering the capital-labour ratio in a way which increases total factor productivity (TFP). In small and low-populated nations among which rank Bahrain, Qatar and Kuwait, ICTs could plug skill gaps in the labour pool by providing capital intensive methods which are more versatile than traditional production techniques. The experience of the Asian countries during the last three decades has shown that technological innovation and skill development made significant contribution to their rapid economic growth. Technical progress strengthens the capability of the economy to generate growth through the creation of new methods and acquiring skills that allow the country to enjoy comparative advantage. For the GCC countries, building of knowledge capacity based on science and technology is critical for the future development of the region. Enhanced cooperation among member states could speed up the process of scientific and technological progress and overcome some of the challenges facing knowledge creation. In other words, the building of strong technological capacity should be put at the top of educational strategy in GCC countries. As pointed out by the United Nations: "Knowledge absorption involves providing people with the capacity to use knowledge via education."[5]

Knowledge acquisition and creation are essential for responding to the challenges facing the GCC countries both internally and externally. Knowledge is treated as a sustainable motor of production linked directly to wealth creation and economic prosperity. Acquiring knowledge is associated not only with productivity growth but also with strengthening the foundation for building human capacity driven by greater tolerance, better communications, and greater understanding. The human capital factor in production depends on knowledge acquisition which exhibits high value-added productive activities needed for promoting rapid economic growth. "A knowledge-based economy is defined as one where knowledge (codified and tacit) is created, acquired, transmitted and used more effectively by enterprises, organizations, individuals and communities for greater economic and social development."[6] Building capacity for promoting such an economy rests heavily on the ability of the nation to manage human resources and to strengthen links to the global educational systems. Meeting these requirements implies that educational policies must encourage diversification and decentralization by allowing universities to set up their own priorities concerning admission, fees, making changes in curriculum and setting rules for quality assurance and achievement awards. In GCC countries, where access by woman to higher education is still constrained by social and cultural factors, educational policy should provide incentives for

---

[4]United Nations, Arab Human Development 2002 (New York: United nations, 2002), p. 7.
[5]United Nations, Ibid, p. 6.
[6]World Bank, Korea and the Knowledge-based Economy (Washington: World bank, 2000), p. 13.

women to enter higher education. Greater participation of women in higher education will tap au underutilized pool of potentially highly skilled labour by increasing the extent of female participation in the new economy. Current labour market rigidities in GCC countries reflect gender inequalities, stemming notably, although not exclusively, from impediments to female student access to science and technology. Universities should be encouraged not only to accept more women students but also should design programs to attract females through scholarships and government support programs. To this end, universities should introduce curriculum that empower domestic industries to hire future knowledge workers, male and female, to take advantages of the opportunities offered by the knowledge-based economy. However, implementation of such strategies depends on the nation's capability to build capacity including setting of monitoring systems to ensure that government policies are in line with the growth of the economy. In addition, universities must carry out scientific research that will be oriented not only towards finding solutions to socio-economic problems but also to increase connection with the global economy. In the Arab world, researchers receive very little recognition for their research reflecting both low public funding and inadequate incentive by the local industry to benefit from research conducted by these researchers. Universities must try to increase contacts with industries in order to tap financial sources of support and to apply the knowledge produced through the research.

5 LABOUR FLEXIBILITY

In contrast to the old economy, the new knowledge-based economy operates on radically different organizational and managerial precepts which require changes in the traditional labour relations. The knowledge worker enjoys greater control over production, operations and technology because of skill-level imparted off-the-job through universities and on-the-job through training *in situ*. However, continuous changes in production techniques and invention of new technologies require an educational system that enables workers to be trained throughout their working life in order to meet desiderata of the new knowledge-based economy. The knowledge worker needs to be flexible by developing specialized skills and gaining knowledge to excel at problem-solving. The new work environment engenders creativity and an approach to work characterized by continuous improvement consonant with quality, which, in turn, requires continuous learning involving upgrading of current skills and acquisition of new.

Perhaps the best way to describe the working life driven by the knowledge-based economy is what the Prophet of Islam said: "seek knowledge from cradle-to-grave". In the new society, the scope of education extends beyond the learning of technical skills and modern knowledge, to include a linkage to the social, economic and human condition. Interdependencies and interconnection among people on a global scale reflects the need to communicate with a much wider audiences of businesses, individuals and groups. With respect to the workforce in the new society, 'the knowledge-based economy requires a skilled, flexible, and adaptable labour force. New developments in the labour market include: rising skill thresholds; falling demand for low-skilled workers in relative terms; and rising growth rates for so-called knowledge workers."[7]

Governments usually intervene in labour markets by imposing regulatory measures that protect local workers and hinder market flexibility. In the new economy, government should increase labour flexibility by allowing workers to adapt to market needs in a highly competitive and rapidly changing global economy. In an open market economy, governments are not in a position to regulate labour markets given that a large portion of demand come from MNCs. These companies select workers on the basis of their skill requirements. Under such circumstances, workers must compete on the basis of their skills and knowledge. The government should contribute to market flexibility by providing special training programs that allow workers update their skills to meet market demand requirements.

---

[7] World Bank, Ibid, p. 60.

Labour market flexibility is facilitated by individual mobility and technical skill specialization in a modern international economy integrated by acceleration in the movement of goods, ideas, technology, information and people. Workers can take advantage of such movement by gaining skills, knowledge and expertise which enhance their ability to secure better job opportunities. Such movement, particularly by workers from developing countries, improves the living standards of workers and their dependents at home. It was estimated that in 2005 about 200 million people, mostly youth, live and work outside their countries. Moreover, a large number of migrants leave homes seeking better opportunities elsewhere. Interestingly, China and India in recent years have been experiencing a reverse 'brain gain' in which rapid ICT growth in these countries created large demand for a return of highly skilled and educated nationals, be they men or women. Changes in work environment, particularly in services, have created enormous demand, in the aftermath of the ICT revolution, for women knowledge-workers. To this end, education and training represent a core prerequisite to build ICT capacity from which women can only be excluded at the peril of any national a development strategy predicated on the use of information and knowledge. Table 2 shows the degree of preparedness for globalization in Arab countries. This table illustrates that with the exception of Dubai, most countries in the Arab world are not ready for globalization. Only Lebanon and Jordan enjoy labour flexibility that qualifies them to make the necessary adjustments in labour markets and increase access to globalization.

Labour markets in GCC countries are greatly influenced by social, cultural, economic, educational, and gender rigidities. Such rigidities usually reduce a country's capability to respond quickly to market changes, especially international requirements for knowledge and skills. As a consequence, GCC country linkage to global markets will remain limited and attempts to gain from globalization will be frustrated. Among the important factors that usually hamper market flexibility are government laws and regulations concerning employment policies, wages, workers benefits, and workers resistance to reforms. Increasing market flexibility requires greater government involvement through training and skill development needed to meet market demand. This

Table 2. Status of GCC states and other Arab countries' requirements for globalization.

Required area	Good	Fair	Poor
Macroeconomic conducive Environment minimal dependency on Volatile source of income	Dubai	Egypt, Jordan, Lebanon	GCC, Iraq, Yemen, Syria
The public sector as regulator	Dubai	None	All
Strengthened private sector	Dubai	Egypt, Jordan, Lebanon	Syria, Iraq, Yemen
Physical infrastructure	GCC	Egypt, Jordan, Lebanon	Syria, Iraq, Yemen
Institutional reform Legal system attracted to FDI	GCC (oil industries)	Egypt, Jordan, Lebanon	Syria, Iraq, Yemen
Increased transparency and curtailed corruption	None	Egypt, Jordan, Lebanon	GCC, Syria, Iraq, Yemen
Competitiveness of labour	None	None	All
Labour productivity amongst nationals	Lebanon, Jordan	Palestine, Syria, Egypt	GCC, Iraq, Yemen
Wage flexibility	None	None	All
Flexible labour markets and skills	Lebanon, Jordan	Egypt	GCC, Syria, Iraq, Yemen

Source: United Nations, Economic and Social Commission for Western Asia, Globalization and Labour Markets in the ESCWA (New York: United Nations, 2001).

will also enhances labour adaptability to the new economy driven by globalization. In a rapidly changing economy, industries need to restructure their operations continuously to keep up with and surpass market requirements. In addition, the government needs to strengthen the social safety net to protect workers from unemployment due to unpredictable changes in market conditions. Globalization is increasing market vulnerability for both products and labour markets due to the rapid changes in technology and uncertainty about global demand and prices. In reference to the labour market in GCC countries, United Nations asseverates that:

> *"Because of the large number of expatriates in the Gulf region and the heavy dependency on them to perform a wide variety of jobs which nationals are reluctant to undertake, labour market in GCC states suffer from a number of structural inflexibilities that vary in intensity from one country to another. Such problems include:*

1. the difficulty of substituting nationals for expatriates
2. labour markets which are made inflexible by labour law restrictions
3. the high wage expectations of nationals
4. the high social and economic costs of maintaining expatriates
5. concentration of nationals in public sector jobs and the increasing number of new entrants to the labour market caused by
   a. demographic factors
   b. increases in the educational attainment of nationals
   c. increasing participation of women in the labour force.[8]

Labour market flexibility in GCC countries is more complex due to market constraints including the cultural and social elements that impose restriction on labour mobility and limit worker freedoms to choose among opportunities offered to them by the new economy. Correcting such imperfections enhances flexibility by making labour more adaptable to changes in global demand as well as by increasing global linkages. The GCC countries need to deregulate and decentralize the labour market by adopting measures that increase mobility within the region and permit access for global markets. Although these countries spend a relatively high share of public expenditure on education, the educational system is still not fully orientated towards promoting a knowledge society. Restructuring education to meet market demand is imperative not only for increasing flexibility but also for enhancing productivity that enables the region to deepen integration into global markets.

In an open economy driven by global forces, the role of ministries of education is expected to become less effective in meeting demand requirements for labour. In the new economy, MNCs could impose their own skill requirements and conduct their own training programmes. Production methods, as well as organizational and managerial competence, require expertise and skills not necessarily available in-country. The role of the ministry of education in promoting the knowledge-based economy will be in the form of making macro-decisions concerning market trends, improving quality, strategic planning, co-ordination and evaluation, and setting benchmarks for meeting global standards. Ministries of Education ought to consign to universities, especially private universities which have the advantage of public universities in terms of the former's close linkages with the private sector, the responsibility to train the future knowledge workforce of the State.

6 THE ROLE OF UNIVERSITIES

Unlike other developing countries, the ability of GCC countries to absorb scientists and engineers through investing in human resources or through recruitment of expatriates could facilitate closing the knowledge gap. In the first case, establishing universities and technical schools is critical

[8] United Nations, Globalization and Labour market in the ESCWA Region, Ibid, p. 43.

for promoting rapid integration. These institutions contribute to the productivity of the economy by undertaking entrepreneurial, managerial, organizational, financial, and leadership activities with the intention of supporting development through increasing skills and knowledge in society. In this vein, the key role of educational institutions, particularly at the tertiary level, including universities, revolves around the imparting of skills to create a knowledge workforce to meet the demands of the emerging and evolving new economies.

Universities can also establish links with business and industries to induce production of high-tech goods and services. "It can conduct R&D for industry; it can create its own spin-off firms; it can be involved in capital formation projects, such as technology parks and business incubator facilities; it can introduce entrepreneurial training into its curricula and encourage students to take research from the university to firms."[9] Unfortunately, in developing countries, including the GCC States, the attitude towards getting a university degree is for family pride and social considerations and not necessarily for its knowledge value. Universities and other educational institutions grant young people degrees to apply for jobs and not to build up knowledge needed for strengthening the productivity of the economy. To assume greater responsibility, universities should be incorporated in the national strategy for social and economic development in order to meet the national skill requirements.

Universities contribute to labour market flexibility by outputting a versatile supply of labour in which graduates are imparted with multifarious technical and scientific skills. In other words, universities ought to induce rapid economic and social development by graduating students with technical skills sufficient to meet demand in labour market instead of acting as degree "mills" in which certificates are awarded irrespective of attained skill level. Universities must make efforts to create links between industry and government. To achieve such, radical measures of academic restructuring need to be undertaken within the university system with a view to (1) streamlining *curricula*, faculty selection, scholarship and (2) adopting incentives for top students, faculty training, and the introduction of technology and new methods in teaching in line with labour market requirements. The government's role in enforcing such requirements could be instrumental through using financial incentives to influence educational institutions programmes with a view to aligning them with the macro-objectives of the economy. In addition, universities should contribute to the knowledge society by introducing programmes and offering courses that strengthen global linkages and give greater access to knowledge produced in the rest of the world. Given global interdependencies, the task of universities is vital not only in promoting scientific and technological development, but also in increasing human understanding worldwide. Universities could become a forum for communication and cultural exchanges, particularly among young people who will eventually be responsible for making future decisions.

Universities could enhance their educational delivery by establishing partnerships with other universities, particularly in industrial countries. Similar partnerships can also be established with non-governmental organizations, private institutions, and other leading educational bodies to increase access to knowledge and teaching methods which are essential for increasing the contribution of universities to social and economic development. Learning from successfully tested methods in education and training could reduce costs as well as enabling universities to play a more productive role in the economy.

New methods in teaching should be introduced to support the country's capacity to absorb knowledge and disseminate information needed to streamline the economic structure and to increase productivity. In line with this objective, ICTs can increase academic choices available to students in terms of the imparting of knowledge by offering a wide range of learning opportunities.

In the GCC countries, the small size of their labour markets, at least in the cases of Bahrain, Qatar, Kuwait, Oman and the UAE, should encourage universities to develop technologies which can be used for building capacity that fosters the creation and growth of small and medium size

---

[9]United Nations, UN Millennium Project, Innovation: Applying Knowledge in development (London: United nations, 2005), p. 39.

enterprises (SMEs) suitable for development of the region. In the GCC states, large-scale projects, rather than SMEs, have commanded the lion's share of government funding and FDI despite the empirical evidence that SMEs would better serve GCC development in the long run. In addition, globalization is changing patterns of demand more in favour of high-tech products and small and medium manufacturing industries. Endorsing such products could be done with the help of governments through public-private partnership, low interest credits, tax incentives, R&D and various other policies that promote global integration.

Universities must also improve the capacity to globalize technologies, especially ICTs. "Globalization of technology falls into three categories: the international exploitation of nationally produced technology, the global generation of innovation, and global technological collaborations"[10] Irrespective of this categorization, universities can serve as incubators for producing technologies that induce building capacity for SMEs. ICTs enable the transference of knowledge through increases in connectivity. Furthermore, conducting research in universities is vital for generating new technologies that can be used in sustaining development. Economic development in GCC countries is constrained by environmental, natural, and water problems which need to be addressed if rational development is to take place. Most funding for R&D in GCC states is not of sufficient critical mass to achieve a technological breakthough able to achieve a ratcheting up in capacity. Rather, making R&D effective in GCC countries requires pooling resources to conduct studies in priority areas that ensure solving some of the problems shared by the people of the region; otherwise, the resources devoted would inevitably be too scant to generate a comparative advantage for the region. In other words, without government support, creating a suitable environment for stimulating indigenous technologies, especially ICTs, would not be possible inasmuch as economies of scale for efficient R&D would not be attained. Endorsing high quality research requires the creation of scientific advisory bodies or agencies with independent budgets capable not only of identifying projects and specific research areas but also collecting data and allocating funding for feasible projects at the GCC, rather than the national, level.

Universities in GCC countries need to increase links with international universities to gain exposure and experience in applying knowledge and teaching. Global education ought to be coopted locally. The GCC countries ought to encourage foreign universities to set up branches and encourage offering joint degrees with local universities. In short, many fundamental changes need to be taking place in GCC countries in order to meet the challenges of globalization. These changes may involve a comprehensive restructuring that include all aspects of the educational system. The new system must meet the need of the knowledge-based economy by creating a suitable environment not only for generating skills and conducting research and development, but also for meeting the needs of GCC citizens for lifelong learning. Table 3 provides statistical information on population and education in selected countries in the Middle East and North Africa. The table indicates that both female participation in the workforce and tertiary enrollment are inadequate to meet the escalating skill demands of globalized labour markets. These two indicators reflect labour market rigidity as well as demonstrate the potential negative impact that a low level of skills could have on global competitiveness and knowledge readiness exacerbated by the absence of greater participation of women in the economy.

In addition to secondary education, higher education in scientific and techno-logical subjects plays an important role in socio-economic development inasmuch as promoting education in the knowledge-based economy requires more than just general education. In addition to knowledge-absorption capacity, knowledge-based economies require knowledge creation. Competition in the global economy depends on production and exports of services and high-tech manufactured goods which contain high quality, technical-skill-intensive labour input. Emphasis should be directed towards science, technology and innovation at the tertiary (as well as the secondary) level. Building adequate human capital capacity emphasizes the importance of adopting *curricula* specifically designed to make science and technology compulsory subjects taught in institutions

[10]United Nations, Ibid, p. 121.

Table 3. Population and education inputs, selected countries, 2004.

Country	Population	% below age 15 (2005)	Female (% of labour force)	Net secondary enrollment ratio	Adult literacy rate (% age 15 and above)	Tertiary enroll-ment as % of total students
Bahrain	0.7	27	–	90	87.7	21
Kuwait	2.6	24	24.8	78	82.9	22
S. Arabia	24.6	37	14.8	52	79.4	28
Oman	2.4	34	15.6	75	74.4	13
UAE	4.6	22	13.2	62	77.3	22
Qatar	0.8	22	–	87	89.2	–
Egypt	74.0	34	21.8	79	55.6	29
Morocco	30.7	31	25.4	35	50.7	11
Jordan	5.8	34	24.1	81	89.9	35
Yemen	20.7	46	27.7	34	49.0	9
Algeria	32.8	30	30.2	66	69.8	20
Syria	18.5	37	30.4	58	82.9	–

Source: United Nations, Human Development Report 2006 (New York: United Nations, 2006); World Bank, World Development Indicators, 2006 (Washington: World Bank, 2006).

of higher learning. In this respect, investment at the secondary and tertiary levels becomes vital in order to increase the number of technical people including those with engineering and scientific degrees.

In this regard, the contribution to development of women who are integrated into the knowledge workforce must be maximized, particularly in non-industrialized countries, where empowerment and employment of women remain inadequate to realize the human capital potential of the national economy. In GCC countries women still lag behind in science and engineering courses which handicaps female participation in the new economy. Among all the female students enrolled in universities in Arab countries, only a very small portion has undertaken specializations associated with science and technology. A change in attitude towards women requires creating a new culture that provides more incentives for women to participate.

Providing women with enabling technical skills and modern knowledge will increase the rate and the value of their participation in the workforce. Beyond the positive impact on productivity of the economy, moreover, increasing workforce participation of women contributes to national welfare and social stability.

Reflecting a positive trend towards women becoming an important segment of the labour market, enrollment of women at various educational levels has registered substantial increases in recent years. In conservative societies such as the GCC countries, the new technologies could provide ample opportunities for the virtual employment of women in which they would work out of home instead of going to an office. In these countries, female participation in the labour force is the lowest in comparison with other regions in the world. In 2004, the share of women in the total labour force was 14.8 per cent in Saudi Arabia, 13.2 per cent in the UAE, 15.6 per cent in Oman, and 24.8 per cent in Kuwait. Unemployment particularly impacts young people and the unemployed rate among the young in those countries may be estimated at double those rates. Among the youth unemployed, women of course feature prominently, especially in Saudi Arabia. "Population growth is adding about six million labour-force entrants every year, a flow that is proportionately greater than in any other region"[11] To this end, the Arab countries need to construct

[11] United Nations, Arab Human Development Report 2002, Ibid, p. 10.

long-term strategies that aim to "move from employment in low-skill, low-productivity sectors to more skill-intensive, high-productivity jobs. Such strategies should exploit the opportunities and niches of globalization."[12]

7 CONCLUSION

The impact of the new global economy on labour markets in GCC countries is pushing the role of local universities in the GCC into the forefront of development. Local universities can address issues of labour inflexibility prevalent in the GCC in relation to managing globalization and facilitating integration in the global markets by imparting technical skills to future national work-force entrants, especially women. In the GCC countries, labour markets are not effective and there is a mis-match between the supply of labour currently leaving educational institutions and the demand of MNCs and SMEs. Owing to artifical constraints imposed by well-intentioned but inefficient governmental regulatory laws, labour markets in the GCC have become hamstrung by market imperfections which inhibit low labour mobility and slow reaction time to market changes. In contradistinction to the lack of dynamism exhibited by otiose GCC labour markets, the new economy is rapidly changing production methods, organizational and managerial techniques and introducing new technologies that need greater labour market response for adjustment. In the new economy the importance of labour and capital has been supplanted by knowledge absorption. Under such circumstances, gaining access to global markets implies that nations must build capacity to strengthen global linkages and enhance competitiveness and a key actor in this transformation is the local university.

Inasmuch as education, especially at the tertiary level, is expected to contribute positively to leapfrogging, by increasing the nation's capabilities to produce skills and develop new products, universities have become instrumental not only by increasing market flexibility and enhancing knowledge absorption, but also by creating new technologies and conducting R&D. Facilitating global linkages and getting access to global knowledge depend on building technological capacity through universities that enable the nation to apply, absorb and eventually create indigenous knowledge. Economic diversification underlines the importance of skills development in generating labour flexibility for managing globalization and enhancing competitiveness. Universities are key strategic resources for governments to plan national skills development for its citizenry.

GCC ministries of education should develop a vision based on future market demand consonant with national plans for socio-economic development. To this end, structural reforms need to be taken in education, especially the correction of gender bias against women which is embodied in the culture of the region. The GCC countries have made substantial efforts towards building a knowledge-based economy by investing in education and by permitting a private university system to emerge.

REFERENCES AND BIBLIOGRAPHY

Al-Roubaie, A. & Al-Zayer, J. 2006. Sustaining Development in the GCC Countries: the Impact of Technology Transfer, *World Review of Entrepreneurship, Management and Sustainable Development*, 2(3).

Al-Roubaie, A. & Al-Zayer, J. 2006. Knowledge Readiness and Sustainable Development in the GCC Countries, *World Development Outlook 2006*.

Asian Development Bank 2006. Knowledge for Development: Priorities for 2006–2008, Manila.

Dervis, K., Bocock, P. & Devlin, J. 1998. International Trade among Arab Countries: Building a Competitive Economic Neighborhood, Unpublished paper presented at the Middle East Institute 52nd Annual Conference, Washington, D.C., 1998.

Drucker, P. 1993. *Post-Capitalist Society*, Oxford: Butterworth-Heinemann.

[12] Ibid, p. 10.

Hislop, D. 2005. *Knowledge Management in Organizations.* Oxford: Oxford University Press.
Makdisi, K. & Cherfane, C.C. 2005. *Arab Region.* International Center for Trade and Sustainable Development. Stockholm.
Organization for Economic Co-operation and Development (OECD) (1996). *The Knowledge-Based Economy,* Paris.
Schacter, M. 2000. Capacity Building: A New Way of Doing Business for Development Assistance Organization. *Policy Brief* No. 6. Institute on Governance. Ottawa.
United Nations, United Nations Environment Programme, Division of Technology, Industry and Economics, Capacity Building on Environment, Trade and Development Trends, Needs and Future Directions, Draft UNEP/ETB 12-07-02.
United Nations 2001a. Economic and Social Commission for Western Asia. *Globalization and Labour Market.* New York.
United Nations 2001b. Potential of Manufacturing SMES for Innovation in Selected ESCWA Countries. New York.
United Nations 2003a. United Nations on Trade and Development, *E-Commerce and Development Report 2003.* New York.
United Nations 2003b. Knowledge Management Methodology: An Empirical Approach Core Sectors in ESCWA Member Countries, New York.
United Nations (2005). Towards an Integrated Knowledge Society in Arab Countries: Strategies and Implementation Modalities. New York.
World Bank 1999. *World Development Report 1998/1999: Knowledge for Development.* Washington.
World Bank 2006. World Development Indicators 2006. Washington.

Higher Education in the Twenty-First Century: Issues and Challenges – Al-Hawaj, Elali & Twizell (eds)
© 2008 Taylor & Francis Group, London, ISBN 978-0-415-48000-0

Accreditation of learning institutions

Ahmed Y. Ali-Mohamed
University of Bahrain, President's Office for Scientific Affairs, Kingdom of Bahrain

ABSTRACT: In this paper we review the definition set up by various accrediting bodies on accreditation.

Thus, accreditation is considered as a certification of academic quality of an institution of higher learning as defined by Wikipedia, 2007a. In its revised edition of characteristics of excellence in higher education, (CHE-MSA, 1994) defined accreditation as the means of self-regulation and peer review adopted by the education community. The accrediting process is intended to strengthen and sustain the quality and integrity of higher education, making it worthy of public confidence and minimizing the scope of external control. The extent to which each educational institution accepts and fulfils the responsibilities inherent in the process is a measure of its concern for freedom and quality in higher education and its commitment to striving for excellence in its endeavours.

On the other hand International Accrediting for Universities (IUAA, 2005) has taken into consideration the recommendations of the principles accepted by the World Conference in Higher Education sponsored by UNESCO (Budapest, 1999) as appropriate for accrediting Universities/School. The long-term outcome of the conference will be efficient and effective renovation and renewal of higher education systems and institutions based in the principles of relevance and quality and with a commitment to enhance international co-operation and academic solidarity. In its clarification IUAA accrediting is a "Third Party" evaluation and "Seal of Approval" for "Quality" educational offerings for the practical training programmes. International Accrediting by IUAA is provided as a free Public Service to the Schools, to students and to the employers of the world. It is also intended to clarify that Accredited Schools are expected to maintain an ongoing programme of self-evaluation to maintain standards of excellence. Schools are expected to consider input and constructing suggestions from faculty, students, parents and the public when developing or maintaining programmes.

In this paper we look at some variation by nations on the accreditation issues of an institution of higher learning and ask if it is mandatory or required by law?

Discussion will be concentrated on the criteria used as a guide to meet the accreditation setup by the accreditation commissions. It will also list the accrediting organizations identified and recognized by the Council for Higher Education Accreditation (CHEA, 2006) in the U.S.A as well as the list of Wikipedia (2007b) of unrecognized accreditation associations of higher learning.

QUESTIONS

1. What do we mean by Accreditation? Can it be done locally, regional or at an international level?
2. What are the criteria needed for accrediting an institute of higher learning?
3. Who is able to accredit an institution of higher learning? Who recognizes the accrediting organizations?
4. Is it voluntary or imposed by some Higher Education authority?

1 INTRODUCTION

An institution of higher learning or education is a community dedicated to the pursuit and dissemination of knowledge, to the study and clarification of values, and to the advancement of the society it serves (CHE-MSA, 1994). To support these goals, institutions of higher education within the Middle States Region in the USA joined together in 1919 to form the Commission on Higher Education of the Middle States Association of Colleges and Schools, a professional association devoted to educational improvement through accreditation (CHE-MSA, 1994). The Association to Advance Collegiate Schools of Business (AACSB) International was founded in 1916 and began its accreditation function with the adoption of the first standards in 1919. Additional standards for programmes in accountancy were adopted in 1980. AACSB International members approved mission linked accreditation standards and the peer-review procession in 1991. In 2003, members approved a revised set of standards that are relevant and applicable to all business programmes globally and which support and encourage excellent in management education world wide (ACCSB, 2007).

In the USA, ABET, Inc., is responsible for the specialized accreditation of educational programmes in applied science, computing, engineering, and technology. ABET accreditation is assurance that a college or university programme meets the quality standards established by the profession for which it prepares its students. For example, an accredited engineering programme must meet the quality standards set by the engineering profession. An accredited computer science programme must meet the quality standards set by the computing profession.

ABET accredits programmes only, not degrees, departments, college, or institutions (ABET, 2007).

Academic accreditation in Switzerland is conducted by the Centre of Accreditation and Quality Assurance of the Swiss Universities (OAQ). The OAQ carries out accreditation procedures for universities as stipulated in Art.7 of the co-operation agreement between the Federal Government and University Cantons on Matters Relating to Universities of 14 December 2000.

The guidelines governing this work were approved by the Swiss University Conference (SUK/CUS) at its meeting on 5 December 2002. The (SUK/CUS) has since approved the first revision of the guidelines (which has been in force since 1 January 2004).

In Switzerland, unlike in other European countries, the accreditation system operates on a voluntary basis. In accordance with the co-operation Agreement between the Federal Government and the University Cantons, accreditation may be granted to public or private institutions or programmes at university level. In line with International regulations and practices, the procedure consists of three stages: a self-evaluation carried out by the University Unit is followed by an external appraisal by an independent group expert. The decision on whether to grant accreditations is then made by the SUK/CUS following a recommendation by the OAQ (OAQ, 2004).

"Universities, while constitutionally autonomous from government in the U.K, are hugely dependent on public funding and therefore are subject to a heavy burden of accountability and audit. However, while audit is an important aspect of quality assurance (QA), it should not be allowed to become the motivation for it. Effective management of quality, understood as the effectiveness of a programme of learning given the needs of the students in the context of a define set of learning out comes, involves bringing assurance, enhancement and audit systems together while recognizing their different functions and rationale" (McGhee, 2003).

However, the role for money in public services and the perceived need to ensure that quality standards are protected in the context of widening participation is not limited o the U.K. (McGhee, 2003).

Smeby and Stensaker (1999) have reviewed the quality systems used in Sweden, Norway, Denmark and Finland and concluded that systems are closely allied to government general strategies and aims for high education rather than to any particular philosophy for what is good academic quality. However, the relationship between government policy and quality systems, and the implications for university autonomy, are not straight forward.

Akoojee (2002) for example has argued that in the case of higher education in South Africa, QA systems should reflect broader socio-political goals; in the case of South Africa, the goals of social redress, justice and equity. Despite the influence of political rather than pedagogical agenda there is evidence that counter to expectations higher education in Europe, while not convergent in terms of organizational structure, is increasingly convergent in terms of QA systems (Rakic, 2001). Since Rakics' review this process has accelerated with, for example, the QAA qualification descriptors being adopted as the core proposal by the post-Bologna conference of European Ministries of Higher Education (van der Wende and Westerheijden, 2002).

Looking across to other regulated sectors it is clear that the challenges and dilemmas facing higher education are not unique in the context of public expectations of standards and quality. Huitema et al. (2002) argue that quality managers in higher education face policy choices similar to those faced by managers of environmental quality. Using Fischer's (1995) framework of policy deliberation, they argue that both sectors face challenges from external stakeholders based around technical verification, situational validation, problem formulation and ideological argumentation".

McGhee (2003) stated that the relationship between public and private sector provision is one that currently does not significantly affect higher education institutions in the U.K. However, McGhee reiterated that "in the context of collaborative provision with overseas partner organization (POs) it is important to recognize the role of QA as a regulatory resource."

He has given an example, Mehrez and Mizrahi (2000) who have highlighted the tensions that occur when government policy incorporates significant private non-university provision but government is reluctant to intervene to monitor threshold standards, either with new institutions or in relation to their impact on standards in the established institutions.

In addition to the rapidly changing economic and regulatory environment in which universities must exist in the 21st century is the unravelling of many of the cultural and ideological constructions on which the very idea of a university is founded in western thought. Established higher education institutions were largely developed in the context of rationalist, scientific and modernist society, and as the certainties of those foundations dissolve with the dawning of a post-modern, 'chaotic' distributed knowledge society, universities must re-examine their role in broader economic and social terms (Scott, 2002; Smith and Webster, 1997).

McGhee (2003) indicated that the future of quality as reported by Pond (2002), writing about the challenges facing traditional US institutions in the light of distributed learning made possible through the Internet, has argued that there is a fundamental mismatch between traditional accrediting paradigms and new educational realities. He suggests that the private sector rather than the government will establish 'consumer-based' means for judging quality, with the possibility of an Amazon or e-Bay for learning programmers.

Alternatively, he proposed alliances of universities, private providers, learners, professional organizations and business will develop their own quality regions. Finally, he suggested the possibility of an accreditation of learner (outcome) achievement rather than recognition of (course) credit or the status of the institutional institution. Despite Pond's assertions to the contrary it seems difficult to see how any 'post-accreditation consortium' could manage QA as distinct from customer preference. However, McGhee (2003) has suggested that the shift from campus-based traditional programmes to distributed and assembled learning packages may accelerate the move towards threshold standards of accreditation underwritten by universities in conjunction with co-ordinating public agencies and customer service-based differentiation managed by private sector bodies.

Finally, the International University Accrediting Association at no charge to the University has fully supported and here-in adopted in full, as appropriate for accredited universities/schools, the recommendations of the world conference on higher education sponsored by UNESCO. The IUAA clearly stated that schools applying for accreditation should be willing to adopt those principles in full. To be accredited, applications must include verifiable evidence of legal authority to grant degrees; incorporation, by itself, does not provide legal authority to grant degrees (IUAA, 2005).

What do we mean by Accreditation? Can it be done locally, regional or on international level?

There are several definitions concerning accreditation. In the USA, it is stated that accreditation is the means of self-regulation and peer review adopted by the educational community. The accrediting process is intended to strengthen and sustain the quality and integrity of higher education, making it worthy of public confidence and minimizing the scope of external control.

The extent to which each educational institution accepts and fulfils the responsibilities inherent in the process is a measure of its concern for freedom and quality in higher education and its commitment to striving for and achieving excellence in its endeavours (CHE-MSA, 1994).

Accreditation assures quality as stated by ABET, 2007. In the United States, accreditation is a non-governmental, peer review process that ensures educational quality.

Educational institutions or programmes volunteer to undergo this review periodically in order to determine if certain criteria are being met. It is important to understand, however, that *accreditation* is not *a ranking system*. It is simply assurance that a programme or institution meets established quality standards.

There are two type of accreditation: Institutional and Specialized.

Institutional accreditation evaluates overall institutional quality. One form of institutional accreditation is regional accreditation of colleges and universities.

Specialized accreditation examines specific programmes of study, rather than an institution as a whole. This type of accreditation is granted to specific programmes at specific levels. Architecture, nursing, law, medicine, and engineering programmes are often evaluated through specialized accreditation.

In the United States, ABET, Inc., is responsible for the specialized accreditation of educational programmes in applied science, computing, engineering, and technology.

Thus ABET accreditation is an assurance that a college or university programme meets the quality standards established by the profession for which it prepares its students. For example, an accredited engineering programme must meet the quality standards set by the engineering profession. An accredited computer science programme must meet the quality standards set by the computing profession.

On that basis, ABET accredits post-secondary degree granting programmes housed within regionally accredited institutions. *ABET accredits programmes only, not degrees, departments, colleges, or institutions.*

AACSB International accreditation assures quality and promotes excellence and continuous improvement in undergraduate and graduate education for business administration and accounting (AACSB, 2007).

Accreditation is a process of voluntary, non-governmental review of educational institutions and programmes. Institutional accreditation reviews entire colleges and universities. Specialized agencies award accreditation for professional programmes and academic units in particular fields of study. As a specialized agency, ACCSB International grants accreditation for undergraduate and graduate business administration and accounting programmes. Institutional accreditation reviews entire Colleges and Universities.

Any appropriately authorized collegiate institution offering degrees in business administration and accounting may volunteer for AACSB International accreditation review.

AACSB International accreditation represents the highest standards of achievement for business schools, worldwide.

Institutions that earn accreditation confirm their commitment to quality and continuous improvement through a rigorous and comprehensive peer preview. AACSB International accreditation is the hallmark of excellence in management education.

It assures stakeholders that business schools:

Manage resources to achieve a vibrant and relevant mission; advance business and management knowledge through faculty scholarship; provide high-calibre teaching of quality and current curricula; cultivate meaningful interaction between students and a qualified faculty; produce graduates who have achieved specified learning goals (AACSB International, 2007).

In the Swiss system, accreditation is defined as a formal and transparent process that uses defined standards to examine whether institutions and/or programmes offered at university level comply with minimum quality requirements.

Accreditation increases the national and international visibility of university performance and can provide students, representatives of universities, politicians, employers and the general public with guidance and an aid in making decisions. The purpose of accreditation is, moreover, to achieve international recognition and to improve the comparability of degrees. In Switzerland, unlike in other European countries, the accreditation system operates on a voluntary basis. In accordance with the co-operation Agreement between the Federal Government and the Cantons, accreditation may be granted to public or private institutions or programmes at university level. In line with international regulations and practices, the procedure consists of three stages: a self evaluation carried out by the university unit is followed by an external appraisal by an independent group of experts. The decision on whether to grant accreditation is then made by the SUK/CUS following a recommendation by the OAQ centre of accreditation and quality assurance of the Swiss Universities.

The decision is based on all the available documentation (self-evaluation report, expert's report, opinion from the university) and may be "yes", "no" or a "yes under conditions". Unconditional accreditation is valid for seven years. Universities which have not yet started operating or which only recently began operating can apply for preliminary accreditation. This expires after three years. Private applicants undergo preliminary evaluation by the OAQ on the basis of a published list of criteria. If the candidate passes the preliminary evaluation, the OAQ proceeds to the accreditation process. If the candidate fails, the SUK/CUS rejects the application for accreditation. The accreditation decisions reached by the SUK/CUS may be submitted to independent arbitration or appeal (OAQ, 2007).

Wikipedia, the free encyclopaedia, (2007a) defines accreditation as a certification of the academic quality of an institution of higher learning. Some countries have independent/private organizations that oversee the school accreditation process, while other countries accredit through a government agency.

Some countries require accreditation and others consider it voluntary. In either case accreditation denotes academic quality and schools that lack recognized accreditation often claim accreditation from unrecognized sources. Unrecognized accreditations are meaningless to the academic community (Wikipedia, 2007b).

The definitive framework for qualification that applies to causes and programmes of study in higher education in England, Wales and Northern Ireland is that published by the QAA in January 2001: the framework for Higher Education Qualifications (FHEQ). The National Committee of Inquiry into Higher Education (NCIHE 1997-the Dearing Report) and Scottish equivalent (the Garrick Report) both recommended the establishment of a qualification framework. This is stated in the following paragraph (McGhee, 2003):

> *"Students needed to be clear about the requirements of the programmes to which they are committed, and about the levels of achievement expected of them. Employers want higher education to be more explicit about what they can expect from candidates for jobs, whether they have worked at sub-degree, degree, or post-graduate level. Existing arrangements for safeguarding standards are insufficiently clear to carry conviction with those who perceive present quality and standards to be unsatisfactory. We believe there is much to be gained by greater explicitness and clarity about standards and the level of achievement required for different awards (NCIHE, 1997: S10.2)."*

The benchmarking academic standards: the subject benchmark statements (SBSs) express expectations in relation to curriculum, skills and standards. In terms of curriculum they lay down a 'broad framework' (rather than any detailed syllabus) for the subject knowledge that an undergraduate degree should contain. In terms of skills, expectations are provided on variously the cognitive, subject and transferable skills that graduate in any given discipline area might be expected or possess. In terms of standards, expectations are expressed as 'threshold' and/or 'typical' student achievement. Threshold means third-class honours degrees and typical means upper-second. In some cases reference in made in the standards sections to 'levels of excellence' meaning 'first-class degree'. McGhee (2003) has focussed on the notes raised by Jackson (2002): QAA benchmarks only incorporate one of three potential elements of any benchmark. They do involve a reference point against which similar programmes can be compared but they do not include a criterion (i.e., a dimension or indicator) against which or alongside which something could be measured, nor gradations of distinction which would mark out the poor, the good, the excellent or the exceptional. The aim of the benchmarks is to provide a common point of reference for achievement of academic standards. However, while the aim of the overall quality infrastructure of the FHEQ, the code of practice and the programme specification system is to provide a context for the broad comparability of academic standards across the sector, the aim of the SBS system is also clearly developmental. The SBS system serves to make both entire academic communities and institutional programme-teams more explicit about what they judge to be key elements of the undergraduate curriculum and to engage in discussion about the boundaries, definitions and priorities that such explicitness renders more conspicuous.

What are the criteria needed for accrediting institute of higher learning?

As mentioned earlier there are two types of accreditation: institutional and specialized. The criteria needed for institutional accreditation are based on the results of an institutional self-study and an evaluation by a team of peers and colleagues assigned by the commission (CHE-MSA, 1994). Accreditation attests the judgment of the commission on Higher Education that an institution has met the following criteria:

a. That it is guided by well-defined and appropriate goals.
b. That is has established conditions and procedures under which its goals can be realized.
c. That it is accomplishing its goals substantially.
d. That it is so organized, staffed and supported that it can expect to continue to accomplish its goals.
e. That it meets the standards of the Middle States Associations Commission on Higher Education.

The same commission has stated standards for accreditation, on the basis that, while the characteristics of accredited educational institutions depend largely on the type of institution, all accredited institutions possess important common attributes. These common characteristics of excellence are the standards by which the Commission on Higher Education determines an institution's accreditation (CHE-MSA, 1994):

1. Integrity in the institution's conduct of all its activities through humane and equitable policies dealing with students, faculty, staff, and other constituencies.
2. Clearly stated mission and goals appropriate to the institution's resources and the needs of its constituents.
3. Clearly stated admissions and other student policies appropriate to the mission, goals, programmes, and resources of the institution.
4. Student services appropriate to the educational, personal, and career needs of the students.
5. Faculty whose professional qualifications are appropriate to the mission and programmes of the institution, who are committed to intellectual and professional development, and who form an adequate core to support the programmes offered.
6. Programmes and courses which develop general intellectual skills such as the ability to form independent judgment, to weigh values, to understand fundamental theory, and to interact effectively in a culturally diverse world.
7. Curricula which provide, emphasize, or rest upon education in arts and sciences, even when they are attuned to professional or occupational requirements.
8. Library/learning resources and services sufficient to support the programmes offered and evidence of their use.
9. Policies and procedures, qualitative and quantitative, as appropriate, which lead to the effective assessment of institutional programmes, and student learning outcomes.
10. Ongoing institutional self-study and planning aimed at increasing the institution's effectiveness.
11. Financial resources sufficient to assure the quality and continuity of the institution's programmes and services.
12. Organization, administration, and governance which facilitate teaching, research, and learning and which foster their improvement within a framework of academic freedom.

13. A governing board actively fulfilling its responsibilities of policy and resource development.
14. Physical facilities that meet the needs of the institution's programmes and functions.
15. Honesty and accuracy in published materials and in public and media relations.
16. Responsiveness to the need for institutional change and renewal appropriate to institutional mission, goals, and resources.

In the United Sates, specialized accreditation examines specific programmes of study, rather than an institution as a whole. Therefore, ABET, Inc., is responsible for the specialized accreditation of educational programmes in applied science, computing, engineering and technology. The quality standards programmes must meet to be ABET-accredited are set by the ABET professions themselves.

This is made possible by the collaborative efforts of many different professional and technical societies. These societies and their members work together through ABET to develop the standards, and they provide the professionals who evaluate the programmes to make sure they met those standards.

General Criteria for Accrediting Applied Science Programmes and Engineering Programmes at BSc levels are:

1. Students: the quality and performance of the students and graduates are important considerations in the evaluation of an applied science programme.
2. Programme educational objectives.
3. Programme outcomes and assessments.
4. Professional component requirements specify subject areas appropriate to applied science programmes, but do not prescribe specific courses. The program's faculty must assure that the applied science curriculum devotes adequate attention and time to each component, consistent with the objectives of the programme and institution.
5. Faculty: the faculty is the heart of any educational programme. The faculty must be of sufficient number as determined by student enrolment and the expected outcome competencies of the programme.
6. Facilities: classrooms, laboratories, and associated equipment must be adequate to accomplish the programme objectives and provide an atmosphere conducive to learning.
7. Institutional support, financial resources and constructive leadership must be adequate to assure the quality and continuity of the applied science programme.
8. Programme criteria: each programme must satisfy applicable programme criteria. Programme criteria provide the specificity needed for interpretation of the basic level criteria as applicable to a given discipline.

However, there are general criteria for Master's level programmes, for technician level Associate Degree Programmes, and for specific disciplines under Applied Science and Engineering Programmes. One can consult (ABET, 2007—Criteria for Accrediting Applied Science Programmes, and ABET, 2007—Criteria for Accrediting Engineering Programmes).

As for the Criteria for Accrediting Computer Science Programmes these are as stated by ABET during the 2007–2008 Accreditation cycle as follows:

1. Objectives and Assessments: the programme has documented, measurable objectives, including expected outcomes for graduates. The programme regularly assesses its progress against its objectives and uses the results of the assessments to identify programme improvements and to modify the programme's objectivities.
2. Student Support: students can complete the programme in a reasonable time. Students have ample opportunity to interact with their instructors. Students are offered timely guidance and advice about the programme's requirements and their career alternatives. Students who graduate the programme meet all programme requirements.
3. Faculty members are currently active in the discipline and have the necessary technical breadth and depth to support a modern Computer Science Programme. There are enough faculty

members to provide continuity and stability, to cover the curriculum reasonably, and allow an approximate mix of teaching and scholarly activity.
4. The curriculum is consistent with the programme's documented objectives. It combines technical requirements with general education requirements and electives to prepare students for a professional career in the computer field, for further study in Computer Science, and for functioning in modern society. The technical requirements include up-to-date coverage of basic and advanced topics in Computer Science as well as an emphasis on Science and Mathematics.
5. Laboratories and Computing Facilities are available, accessible, and adequately supported to enable students to complete their course work and to support faculty teaching needs and scholarly activities.
6. The Institution's support for the programme and the financial resources available to the programme are sufficient to provide an environment in which the programme can achieve its objectives. Support and resources are sufficient to provide assurance that the programme will retain its strength throughout the period of accreditation.
7. Institutional facilities including the library, other electronic information retrieval systems, computer networks, classrooms, and offices are adequate to support the objectives of the programme.

As far as Business and Accounting Accreditation Standards are concerned, the AACSB International Accreditation assures quality and promotes excellence and continuous improvement in undergraduate and graduate education for business administration and accounting for detail information on eligibility procedures and standards for both business and accounting accreditation. The criteria adopted in April 2003 and revised on 31 January 2007 for business, and the criteria adopted in April 2004 and revised on 31 January 2007 for accounting, can be consulted at the webpage http://www.aacsb.edu/accreditation/standards.asp of ACCSB International.

The UNESCO-CEPES (European Centre for Higher Education) has set regulations concerning accreditation, evaluation and approval pursuant to the act relating to universities and colleges and the act relating to private colleges. Accreditation *as a college* depends on the following conditions being met:

a. The institution's primary activities must be higher education, research and dissemination of knowledge.
b. The institution must have the right to award lower degrees for at least one study programme and have awarded lower degrees for at least two years.
c. The institution must have research and development (R&D) related to this field.
d. The institution must have independent competence in major fields included in the study programme.
e. The institution must have an academic library.
f. The institution must have a board with representatives from the staff and students.
g. The institution must have an infrastructure and organization for conducting higher education.

In order to be accredited as a *specialized institution* at *university level*, the following conditions must be fulfilled:

a. The condition stated under accreditation as a college must be fulfilled.
b. The institution must have an academic staff and stable research activities of a high standard.
c. The institution must have an organization and infrastructure for conducting higher education and research.
d. The institution must have the right to award higher degrees for at least one study programme or other programmes of at least five year's duration, and have awarded higher degree for at least two years.
e. The institution must have an independent right to award doctorates and have stable research training.
f. The institution must be affiliated to national and international networks in connection with higher education and research and participate in the national co-operation as regards research training.

In order to be accredited as *University*, the following conditions must be fulfilled:

a. The conditions stated under accreditation as a college must be fulfilled.
b. The institution must have an academic staff and stable research activities of a high standard.
c. The institution must have an organization and infrastructure for conducting higher education and research.
d. The institution must have an independent right to award higher degrees or hold other study programmes of at least five years duration in at least five areas as well as lower degree programmes in a number of fields. The institution must award degrees in three fields.
e. The institution must have an independent right to award doctorates in at least four fields, of which two fields must be of major relevance for regional development while also having national importance. The institution must have stable research training.
f. The institution must be affiliated with national and internationals network in connection with higher education and research and participate in national co-operation as regards research training.

For details on this matter refer to the web-site of UNESCO-CEPES http://www.cepes.ro/cepes/mission.htm

Who is able accredit an institution of higher learning? Who recognizes the accrediting organizations?

In the USA, institutional accreditation of higher learning is accredited either regionally or nationally by accrediting agencies. These are:

- Accrediting Council for Independent Colleges and Schools,
- Distance Education and Training Council,
- Middle States Association of Colleges and Schools,
- New England Association of Schools and Colleges,
- North Central Association of Colleges and Schools,
- Northwest Association of School and Colleges,
- Southern Association of Colleges and Schools,
- Western Association of Schools and Colleges.

The above accrediting bodies are recognized by the United States Department of Education (see the web-site http://www.aju.edu/usdoe_accreditation.htm).

Specialized accreditation is the accrediting programmes of study, rather than an institution as a whole. There are forty-seven specialized and professional accrediting organizations in the USA; some of them are listed below:

- AACSB International—The Association to Advance Collegiate Schools of Business (AACSB),
- Accrediting Board for Engineering and Technology (ABET),
- Accrediting Council for Pharmacy Education (ACPE),
- Accreditation Review Commissions on Education for the Physician Assistant, Inc (ARC-PA),
- Accrediting Council on Education in Journalism and Mass Communications (ACEJMC),
- American Council for Construction Education (ACCE),
- American Dietetic Association,
- Commission on Accreditation for Dietetics Education (CADE-ADA),
- American Library Association (ALA),
- Committee on Accreditation (COA),
- Teacher Education Accreditation Council (TEAC),
- American Occupational Therapy Association (AOTA),
- Accreditation Council for Occupational Therapy Education (ACDTE),
- American Physical Therapy Association (APTA),
- Commission on Accreditation in Physical Therapy Education (CAPTE),
- American Speech-Language-Hearing Association (ASHA),

- Council on Academic Accreditation in Audiology and Speech-Language Pathology,
- American Society of Landscape Architects (ASLA),
- Landscape Architectural Accreditation Board (LAAB),
- Association of Collegiate Business Schools and Programs (ACBSP),
- Commission on Accreditation of Allied Health Education Programs (CAAHEP),
- Commission on Collegiate Nursing Education (CCNE),
- Council on Interior Design Accreditation.

The aforementioned accrediting organizations, among others, are recognized by the Council for Higher Educational Accreditation (CHEA). Recognition by CHEA affirms that the standards and processes of the accrediting organization are consistent with academic quality, improvement and accountability expectations that CHEA has established, including the eligibility standards that the majority of institutions or programs each accredits are degree-granting. For the complete directory of CHEA Recognized Organizations 2006–2007 refer to the web-site http://www.chea.org/Directories/special.asp.

There are forty-five accrediting bodies not recognized by the United States Department of Education; for details refer to the web-site http://www.aju.edu/usdoe_accreitation.htm.

Wikipedia, the free encyclopaedia, (2007b) reported that schools that lack legitimate accreditation often claim accreditation from an unrecognized source. It continues to state that in the United States of America, accreditation from agencies that do not have recognition from the United States Department of Education or the Council for Higher Education Accreditation is "bogus" to the academic community. It also lists ninety-six bodies of unrecognized accreditation association of higher learning; for derails of the list, references and sources, refer to the web-site http://en.wikipedia.org/wiki/list_of_unrecognized_accreditation_association_of_higher_learning.

Wikikpedia (2007a) also reports that some countries have independent/private organizations that oversee the school accreditation process, while other countries accredit through a government agency.

International recognition may be done by IUAA and VUAA who are recognized as International Accrediting Association and/or listed in the data bases of:

UNESCO-CEPES: UNESCO is the education arm of the United Nations; UNESCO-CEPES is recognized by the U.S. Department of Education;

UNESCO is also recognized by the American Council on Higher Education Accreditation (CHEA)—see the Lisbon Convention. IUAA and VUAA are recognized by the European Commission, the Council of Europe and UNESCO/CEPES;

UNESCO/CEPES are recognized by the Council on Education Accreditation (CHEA)—see Lisbon Convention. IUAA and VUAA are also recognized by the Russian International Academy of Science of Information, Communication, Control in Engineering, Nature, Society and Management of Technology, St Petersburg, Russia. They are also recognized by the International Inter-Academic Union in Moscow—recognized by the United Nations and many ministries of education;

IUAA stated that as appropriate for accredited universities/schools, the recommendations of the World Conference on Higher Education sponsored by UNESCO. Accredited Schools are expected to maintain an ongoing programme of self-evaluation to maintain standards of excellence. School are expected to consider input and constructive suggestions from faculty, students, parents and the public when developing or maintaining programmes;

IUAA and UVAA's stated mission is to provide solicited and unsolicited evaluations and accreditations to professional associations and to worthy education programmes from education providers worldwide. They both share UNESCO's belief that there are many forms of higher education. IUAA and VUAA are independent and free to evaluate each educational institution and programme without influence from any source. They will not accept donations or fees as part of an accreditation application;

They also provide services to schools without any fees or charges. Evaluation and accreditation services are free.

IUAA and VUAA stated their international code of ethics as criteria for accreditation:

Provide and excellent education; require excellence from faculty; expect excellence from students; publish a refund policy; clearly state all course requirements; work to keep the tuition and costs affordable; provide an environment of academic freedom; establish non-discriminatory admissions standards; require that teaching and grading be done by professors; allow equal access to all resources to every student; use technology to enhance the educational experience; all accredited schools will give fair consideration to all applicants and to all transferring students. For details, refer to the IUAA web-site http://www.accrediting.com/

2 IS ACCREDITATION VOLUNTARY OR IMPOSED BY HIGHER EDUCATION AUTHORITIES?

According to Wikipedia, the free encyclopaedia, (2007a) accreditation is a certification of the academic quality of an institution of higher learning. Some countries have independent/private organizations that oversee the school accreditation process, while other countries accredit through a government agency.

Some countries require accreditation and others consider it voluntary. In either case accreditation denotes academic quality and schools that lack recognized accreditation often claim accreditation from unrecognized sources. Such unrecognized accreditations are meaningless to the academic community.

Examples of such recognized accreditation associations of higher learning include:

1. In Hong Kong a school does not have to be accredited, but "it is up to your employer or institution to recognize your qualifications". Currently, accreditation in Hong Kong and China are jointly recognized. For details refer to the web-site, http://www.hkcaa.edu.hk/First.htm;
2. In India, accreditations for universities are required by law unless it was created through an act of Parliament. Without accreditation "It is emphasized that these fake institutions have no legal entity to call themselves as University/*Vishwvidyalaya* and to award "degrees" which are not treated as valid for academic employment purposes. For details refer to http://www.education,nic.in/htmlweb/he-centraluniversities-list.htm
   Accreditation for higher learning is overseen by autonomous institutions established by the University Grants Commission. For details refer to the web-site http://www.education.nic.in/higedu.asp;
3. Pakistan: in 2003, Canada began helping Pakistan develop an accreditation system. Currently, these accreditors are recognized by the Higher Education Commission. For detail refer to the site http://www.paskistanlink.com/Headlines/Jun05/02/03.htm;
4. In the United Kingdom the institution-offering degree must be accredited and a list maintained by the Department for Education and Skills. For the official list of genuine education and training providers refer to http://www.dfes.gov.uk/providersregister/;
5. In the United States of America, the United States Department of Education does not directly accredit educational institutions and/or programmes. However, the U.S. Secretary of Education is required by law to publish a list of nationally recognized accrediting agencies in the United States that the Secretary determines to be reliable authorities as to the quality of education or training provided by the institutions of higher education and the higher education programmes they accredit, within the meaning of the Higher Education Act of 1965, as amended.
6. In Switzerland, unlike in other European countries, the accreditation system operates on a voluntary basis. In accordance with the co-operation agreement between the Federal Government and the Cantons, accreditation may be granted to public or private institutions or programmes at University level. In line with international regulations and practices, the procedure consists of three stages: a self-evaluation carried out by the university unit is followed by an external appraisal by an independent group of experts. The decision on whether to grant accreditation

is made by the SUK/CUS following a recommendation by OAQ (Centre of Accreditation and Quality Assurance of the Swiss Universities).

3 CONCLUSIONS

Accreditation of higher learning institutions is the means of self-regulation (self-evaluation) and peer review adopted by the educational community. Accreditation systems operate on a voluntary basis or are imposed by a high educational authority.

Accreditation is a certification of the academic quality of an institution of higher learning.

In certain countries (USA) the Council for Higher Education Accreditation serves students and their families, colleges and universities, sponsoring bodies, governments, and employers by promoting academic quality through formal recognition of Higher Education Accrediting bodies, and co-ordinate and work to advance self-regulation through accreditation.

REFERENCES

ABET—Applied Science Accreditation Commission, Computing Accreditation Commission, Engineering Accreditation Commission, Technology Accreditation Commission-Accreditation Policy and Procedure Manual. Effective for Evaluations During the 2007–2008 Accreditation Cycle.

ABET—Applied Science Accreditation Commission. Criteria for Accrediting Applied Science Programs Effective for Evaluations During the 2007–2008 Accreditation Cycle.

ABET—Computing Accreditation Commission. Criteria for Accrediting Computing Programs. Effective for Evaluations During the 2007–2008 Accreditation Cycle.

ABET—Engineering Accreditation Commission. Criteria for Accrediting Engineering Programs. Effective for Evaluations During the 2007–2008 Accreditation Cycle.

Accreditation Board for Engineering and Technology, Inc. ABET—The Basics Accreditation: http://www.abet.org/the _basics.shtml.

Accrediting bodies NOT recognized by the United States Department of Education. http://www.aju.edu/usdoe_accreditation.htm.

Akoojee, S. 2002. Access and quality in South African Higher Education: The challenge for transformation, HERDSA Conference Proceedings.

Association to Advance Collegiate Schools of Business AACSB International Accreditation: http://www.aacsb.edu.accreditation/.

Association to Advance Collegiate Schools of Business AACSB International Accreditation Standards: http://www.aacsb.edu/accreditation/atandards.asp.

Centre of Accreditation and Quality Assurance of the Swiss Universities (OAQ): http://www.oaq.ch/pub/en/03_01_00_akkredit_hochschul.php

Characteristics of Excellence in Higher Education—Standards for Accreditation. Commission on Higher Education—Middle States Association of Colleges and Schools (CHE-MSA), 1994. Philadelphia, PA.

Council for Higher Education Accreditation (CHEA), 2006–2007 Directory of CHEA Recognized Organizations—Updated 11 October 2006, U.S.A.

Council for Higher Education Accreditation (CHEA), Specialized and Professional Accrediting Organizations (last modified on 8 March 2007): http://www.chea.org/Directories/specila.asp.

Guidelines for Academic Accreditation in Switzerland of 16 October 2003. The Swiss University Conference (SUK/CUS).

Huitema, D., Jeliazkova, M. & Westerheijden, D.F. 2002. Phases, levels and circles in policy development, the case of higher Educational environmental quality assurance. *Higher Educational Policy*, 15: 197–215.

International University Accrediting Association and Virtual University Accrediting Association. http://www.accrediting.com/.

Jackson, N. 2002. Growing Knowledge about QAA subject benchmarking, Quality Assurance in Education, 3, pp 139–54.

McGhee, P. 2003. *The Academic Quality Handbook—Enhancing Higher Education in Universities and Further Education Colleges*. London and Sterling, VA: Kogan Page.

Mehrez, A. & Mizrahi, S. 2000. Quality requirements in rapidly growing higher education systems: The Israeli example. *Higher Education Policy*, 13: 151–171.
NCIHE 1997. Higher Education in Learning Society (The Dearing Report), NCIHE, London.
Pond, W.K. 2002. Twenty-first Century education and training: Implications for quality assurance. *Internet and Higher Education*, 4: 185–92.
Rakic, V. 2001. Converge or not converge: The European Union and higher education policies in the Netherlands, Belgium/Flanders and Germany. *Higher Education Policy*, 14: 225–240.
Regulation concerning accreditation, evaluation and approval pursuant to the act relating to universities and colleges and the act relating to private colleges 2003. Chapter 3: Accreditation of Institutions and Course Programs and Approval of Examinations and Degrees. (PDF format—0.08 MB). http://www.cepes.ro/hed/policy/legislation/he_laws.htm.
Scott, P. 2002. The future of general education in mass higher education. *Higher Education Policy*, 15: 61–75.
Smeby, J.-C. & Stensaker, B. 1999. National Quality Assessment Systems in Nordic countries: Developing a balance between external and internal needs. *Higher Education Policy*, 12: 3–14.
Smith, A. & Webster, F., eds, 1997. *The Postmodern University? Contested Visions of Higher Education in Society*. SRHE and Milton Keynes, U.K.: The Open University.
UNESCO-CESPES (European Centre for Higher Education)—Mission http://www.cepes.ro/cepes/mission.htm.
Van der Wende, M. & Westerheijden, D. 2002. Report of the Conference "Working on the European Dimension of Quality" of the Joint Quality Initiative, Amsterdam, 12–13 March 2002.
Wikipedia, the free encyclopaedia, 2007a. List of recognized accreditation associations of higher learning http://en.wikipedia.org/wiki/List_of_recognized_accreditation_associations_of_higher_learning. (This page was last modified 20:25, 2 March 2007).
Wikipedia, the free encyclopaedia, 2007b. List of unrecognized accreditation associations of higher learning http://en.wikipedia.org/wiki/list_of_unrecognized_accreditation_association_of_higher_learning. (This page was last modified 13:58, 12 March 2007).

Higher Education in the Twenty-First Century: Issues and Challenges – Al-Hawaj, Elali & Twizell (eds)
© 2008 Taylor & Francis Group, London, ISBN 978-0-415-48000-0

Educational priorities for the twenty-first century: Multi-cultural challenging strategies for the amelioration of higher education in Bahrain

Samia Costandi
The Royal University for Women, Faculty of Education, Kingdom of Bahrain

Today, higher education faces strenuous pressure. In our so-called global village, modernity and post-modernity opened the door for major advances in technological communication between peoples across national and cultural boundaries. It is the thesis of this paper that giant strides in technological advancement have *not* been met with commensurate advancements within the teaching and learning of adults, the practice of ethics, the pedagogy of emotional intelligence, or with advancements in leadership and global citizenship—to name a few problematic areas. In short, there is a dearth of real progress in the lives of citizens around the globe that disables them from living the best lives they can. The proof is all around us: rampant wars, hunger, unequal opportunities, poverty of women and children in particular.

I am proposing a framework for higher education that will include the following interdependent and connected aspects:

1. Imaginative thinking at the centre of the framework.
2. Co-operation between the public and the private sectors: dialogue between school and society, and between educators, administrators and politicians.
3. The circle of life: hierarchical and patriarchal paradigms have not worked in education in the past. Women's human, civil, and other rights are not against but within the Arabic Islamic tradition.
4. Modelling, dialogue, practice, and confirmation: encouraging more women to join the work force and become leaders in society; example: Shaikhah Haya Rashed Al Khalifa.
5. Re-conceptualizing globalization: a cosmopolitan global democracy and a sense of great pride in one's local culture.
6. Multi-cultural education makes everyone visible: no-one is invisible.
7. Child-friendly budgets and protective environmental projects by multinational companies: creating non-governmental organizations that can help government by mentoring, offering consultancy services, and monitoring the infractions of multi-national giants.
8. The quality of education that universities in Bahrain offer and the criteria they fulfil—rather than the number of universities—determines the quality of higher education in Bahrain.
9. Nurturing an understanding of emotional intelligence: encouraging multiple paradigms of creative thought in multiple fields for a diversity of careers since learners have multiple intelligences.
10. Make technology succumb to our dreams and serve our aspirations rather than control us.

This paper will highlight salient issues within multi-cultural education which if not addressed will further deepen the rift between school and society. The paper will explore educational venues that may help governments achieve balance between what is perceived as pragmatic economic priorities and, on the other hand, educational priorities—as seen by an academic. The paper seeks to encourage educators and administrators to push the boundaries of intellectual, humanitarian, and emotional growth within the private and public space. The paper provides a framework through which such thinking about collaboration between private and public space in the field of education could be integrated within the milieu of the Gulf and specifically of Bahrain.

1 IMAGINATIVE THINKING AT THE CENTRE OF THE FRAMEWORK

At the outset, I would like us to think about the word "imagination." "Imagination" is a key word in education. I will focus on this word as an introduction to the framework I am suggesting in relation to new ways of thinking about higher education in the Kingdom of Bahrain. I espouse the framework of an *Imaginative Labour of Care* in this conference that I believe could lead to reforms where everyone benefits and no one loses. Most of all, as an educator, I would like to count on *the co-operation of the private and the public sectors* to achieve the goals that educators hope to achieve: namely, creating a reform plan that is so effective that Bahrain becomes one of the leading Gulf countries in the realm of higher education.

The word "imaginative" is commonly used as a synonym for "creative," "sensitive," "non-traditional," and "inventive." I am using Robin Barrow's definition (Professor of Philosophy at Simon Fraser University in Western Canada) and keynote speaker at the recent First Pacific-Rim Conference on Education in 2006: *Emergent Issues and Challenges in Education* (October 21 & 22, Sapporo, Japan). In an article years ago on *the concept of the imagination* (in Egan & Nadaner, eds, 1988), Barrows described the noun, 'imagination,' as being a superior capacity for abstract thinking that fulfils certain criteria and qualifications. It is not any capacity for abstract thinking, nor is it simply a superior capacity, but it is *the quality of conceiving* that makes someone imaginative or not. Imagination, for Barrows, is not a physiological, neural, or chemical part of our brain; rather, it is *the quality of conceiving* real or imaginary situations in a particular way within a certain context (1988, pp. 84–88) that makes an educator, thinker, writer, artist, journalist, physician, engineer, architect, business person, leader, politician, or technician imaginative. Imagination is not merely a preoccupation with the "abstract" as in one's preoccupation with imaginary beings and worlds (as in phantasy) (pp. 80–84); the two universal criteria that make that superior capacity for abstract conceptualization *imaginative* are, according to Barrows, the criterion of "unusual" and the criterion of "effective." (p. 84)

Our work in this conference can be an imaginative labour of love if we are able to produce together a framework for higher education in Bahrain that is both "unusual" and "effective." We may need to explore alternative ways of seeing and doing that will lead to effective results. This labour of ours needs to be characterized by insight, subtlety, farsightedness, and an ability to go beyond the routinely and immediately apparent. Educators, like heroes, need to look beyond the blind spot of humanity, as Joseph Campbell used to say (Campbell 1973, 1988).

2 CO-OPERATION BETWEEN THE PUBLIC AND PRIVATE SECTORS: DIALOGUE BETWEEN SCHOOL AND SOCIETY, BETWEEN EDUCATORS, ADMINISTRATORS AND POLITICIANS

Am I speaking here about learning to do more activities? I do not think so. Am I speaking about enhancing the power of neurons within a certain part of one's brain? Not really. Am I talking about amassing the largest amounts of information in the most modern ways of e-learning or technologically advanced learning? No. I am not speaking about cluttering our minds or desks with yet more technology; I am not even speaking about elevating the rates at which young people are procuring Bachelors, Masters, or Doctoral degrees. I am speaking about *the constant effort to look for unusual and effective ways to bring society and school together*. John Dewey, the most prominent North American educator, emphasized this relationship between school and society as crucial to the success of our educational endeavors (Dewey, 1902/1990). The thesis of my argument here is that if there is no dialogue between school and society, then education is for naught!

Education itself is about developing the quality of the imagination, albeit indirectly, in order to teach certain kinds of curricula in particular ways. Education is about having an experience of learning that is unforgettable and that may give one insights into the amelioration of many areas of life. Education is about a healthy and open sharing of experiences with other teachers, students and learners in order to create the kind of consensus or lack thereof, heated debates and momentum that will drive society forward. Not all change is good, and not all modern ways of teaching and learning are good. We need to find what befits us, what is effective in our societies. We need to find what is befitting for us as humans and for our environment, to construct our own values (Maguire & Fargnoli, 1991). We need to have an ongoing conversation between school and society (Dewey, 1902/1990), between administrators, ministries, teachers, politicians and parents. Our educational philosophy at the Royal University for Women (RUW) focusses on three aspects: teaching, research, and work with the communities.

3 THE CIRCLE OF LIFE: HIERARCHICAL AND PATRIARCHAL PARADIGMS HAVE NOT WORKED IN EDUCATION IN THE PAST. WOMEN'S HUMAN, CIVIL, AND OTHER RIGHTS ARE NOT AGAINST BUT WITHIN THE ARABIC ISLAMIC TRADITION

The concept map I was hoping to present here has to do with bringing together different sectors within society around a paradigm of CARE as Nel Noddings suggests (Noddings, 1992), and under an umbrella of *inter-connectedness*. The web of life, as described by the physicist Fritjof Capra (Capra, 1991) cannot elude us anymore. If we presume to know anything about teaching and learning, then we probably know today that patriarchal and hierarchical paradigms in life and in learning do not work anymore.

I espouse the circle as a symbol of life and not the triangle. I believe we need to create an ecologically viable society, to engage in learning that is fun and is interesting; we need to create possibilities in the minds of the young of their becoming not only active members within society but agents of change; to inspire and empower students. No matter how many skills you teach students, if one does not inspire them to be imaginative thinkers, or instil in them curiosity of the mind and courage of the heart, they will not be able to conceive of change in an unusual and effective way. Skills do not create paradigm shifts; *imaginative thinking does*. Newton's laws were not something that he discovered that were hovering somewhere and had eluded thinkers before him. Newton's laws were in reality *Newton's way of conceiving the world*, Professor Tarnas informs us (Tarnas, 1991). Every philosopher, thinker, and educator has a way of conceiving the world. Those ways that are most imaginative are the ones which eventually create paradigm shifts. Only imaginative thinking will transform society and carry it into equity and egalitarianism among its citizens, into a kind of wellness of being that will make families and communities thrive.

Allow me to tell you what I imagine for my students (our practising teachers) in future as I continue to teach and learn with them at the Faculty of Education at the RUW. I want my young female students to be able to work in multiple fields of their choice in a pluralistic society that is open and in constant dialogue. I want our practising teachers to be empowered enough to not only push for systemic reform, but be able to defend their ideas with good arguments and good examples. I want them to become *critical thinkers*.

Paulo Freire (Freire, 1970), the great Latin American educator, was in constant dialogue with the peasants in his country and in other societies because he believed that everyone can be involved in learning; it is not specific to schooling. Moreover, transformative learning and education for Freire was about empowering the most vulnerable sectors of society. For Freire, students can no longer be passive recipients who merely regurgitate on an exam what the teacher has poured into their so-called passive minds! They need to construct their education together with their teachers. The Constructivist paradigm in educational psychology promotes the same ideas (which by the way were proposed by John Dewey more than a century ago): learning by doing; learning through experience; learning through a hands-on methodology; being a participant in one's own learning; learning through focussing on students' interests. I want my students to be active learners and citizens, to have confidence and have access to different government ministries, offices, educational and social institutions and organizations in order to make suggestions about reform to those in leadership positions within those institutions. I would like them to feel that they can engage in democratic dialogue at all times. I want my students to become leaders themselves in their diverse communities and in society at large in Bahrain.

We need philosophical paradigms in education that focus on interdependence, inter-connectedness, a holistic view of learning, not hierarchical and patriarchal thinking that centralizes power in the hands of people who do not understand education and end up eclipsing hopes and dreams.

4 MODELLING, DIALOGUE, PRACTICE AND CONFIRMATION: ENCOURAGING MORE WOMEN TO JOIN THE WORKFORCE AND BECOME LEADERS IN SOCIETY; EXAMPLE: SHAIKHAH HAYA RASHED AL KHALIFA

We need to recognize the voices and values of Bahraini educators, particularly women thinkers. It is time to listen to women and take their suggestions very seriously in higher education. If we look at the example of Shaikhah Haya Rashed Al Khalifa, who is heading the United Nations sessions today, Bahrainis can really be very proud.

Nel Noddings' framework of CARE in education is one that I strongly encourage (Noddings, 1992). Modelling, dialogue, practice, and confirmation are key words in her paradigm of CARE. Shaikhah Haya Al Khalifa is a perfect example of a woman who has been modelling, has been in dialogue, is practising, and is confirming to other women what needs to be done. She is a great educator, for education does not only take place in schools and classrooms; learning takes place everywhere in both the private and public spaces, at home and outside of it, at work and with peers. The female philosopher Nel Noddings believes that schools have not reacted well in the past to social change: she means that schools continue to emphasize competition and achievement at the expense of caring for the individual student and the whole person. Remember the person who brought you forth into this world: your mother. Remember that she is your first educator. Remember her strength, her wisdom, her groundedness, her unpaid labour, her unrelenting resolve, her uncompromising morals, her constant modelling, her dialogue with her children, her unrelenting practice in the home, and her confirmation of your labour which encouraged you to go forward in life. Let us look at our mothers and learn from them how to be leaders.

Fighting for the rights of women is not against Arab Islamic values: it is at the core of Koranic teachings and has been emphasized in Islam. The Higher Council For Women in Bahrain needs to be given more power to not only act as a referral centre for women's issues and problems; rather, it needs to be mandated to make decisions regarding women's issues; the Council needs to explore problems, provide concrete suggestions and argue for solutions, and even suggest the creation of specific laws that guard the rights of women in the Kingdom. It is a Council for Women and therefore, I believe, understands women's issues best. Societies that have given executive powers to women's institutions have moved faster than others in the areas of reform within higher education.

5 RE-CONCEPTUALIZING GLOBALIZATION: A COSMOPOLITAN GLOBAL DEMOCRACY AND A SENSE OF GREAT PRIDE IN ONE'S LOCAL CULTURE

If you can imagine the world as Mother and Globalization as the Father, you will be able to interpret what is happening to our globe with more insight. The fierce and greedy business-driven kind of globalization which is spearheaded by multi-national companies that are so hungry for profit at the expense of human blood is what we are experiencing today; it is at the root cause of the problems we face in our developing societies. Have you watched the film: *An Inconvenient*

Truth? Look around you: greedy capitalist globalization has created sectarian wars to guard its interests, has uprooted communities and cultures, and has destroyed, swallowed, or bought small companies and ravaged communities. Globalization has not been responsive to public needs in societies which are volatile because they are struggling to emerge slowly from neo-colonialism and tread cautiously on new grounds. This form of globalization we are experiencing today compromises women and children the most.

There is a huge leap. The Arab Middle East is attempting to go through *modernity, post-modernity,* and a *post-partum* of both phases in one fell swoop! It is almost impossible to do that. Hence, David Held's words ring true in one of his interviews: "We need to reconfigure democracy in our global village in order to create protective walls around communities in third world countries that have been ravaged by this mean-spirited neo-liberal project of globalization that is reinforced by the conditionality programmes of the IMF and the World Bank." http://www.globalpolicy.org/globaliz/define/2004/04heldinterview.htm

Why is it, Held asks, that the European countries who built protective walls around their own economies while they were growing, post World War II, are not granting developing countries the same protection that they gave themselves? Why do they push the IMF and the World Bank to impose conditions on developing countries that are unbearable and destructive economically and socially? We need to be included in the paradigm of human rights and democracy that European and North American countries themselves use to guard the rights of their own citizens.

Let us not kid ourselves; the Kingdom of Bahrain is a small, lovely island that does not have the oil resources that other Gulf countries have. It is forced to act as a service country in order to survive. There is nothing wrong with Bahrain being a country of services, for example financial services, tourist services, etc. We need to *protect* this jewel of an island fiercely, however, if we are to achieve at least some of the aims of higher education here. One of the important aims of higher education is to guard the fabric of society, to build a strong infrastructure, to maintain the web of life within the indigenous spaces. We have the full right to create laws that guard Bahraini interests. Do not believe it when the hierarchical global voice tells you this will drive companies away! There is enough existing competition and indigenous populations must have choices. Indigenous populations can create laws that guard their interests. There are so many companies that continue to make infractions, to be greedy and unfair to employees and totally unscrupulous in their behaviour and dealings just because, simply because, they can get away with it!

In our old yet young societies (Delmon is 5000 years old but Bahrain is fairly young) we need to be able to be part of the world community's current financial and business activities without compromising our national heritage, our national identity, and our Arab Islamic civilization and values. The great late comparative literary critic, Edward Said, scholar, aesthete and political activist, has given us many examples of how our Arabic Islamic civilization has been compromised in the media; how we have become victims of stereotyping and denigration consistently. Said's most famous book, *Orientalism*: *Western conceptions of the Orient* (Said, 1978) turned the tables on Western orientalist scholarship. Why are our students not studying his works in universities in the Gulf today? We need to use the books by authors like Edward Said (1978, 1993, 1994, 1997), Issa Boullata (1990), Albert Hourani (1991), Naiim Ateek (1989), and like Ameen Maalouf (1984) (who discussed the Crusades through Arab eyes) to open the minds and hearts of students and adult learners in our society to be proud of our national heritage. We need to learn about Arab culture, civilization, literature, philosophy, contributions of our great civilization to world civilization; we need to have *a humanistic renaissance*, so to say, of Arab culture and learning.

In our Arab Middle East world of today, the only way our students are going to succeed is by having a strong sense of national and individual identity, to be proud of themselves and not only to emulate other cultures. Incidentally, all the authors I mentioned above are Christian Arabs; there is a whole legacy there that the average Arab citizen does not necessarily know, which is that some of the most prolific and eminent scholars of Arab Islamic civilization and heritage are Christian Arabs. What does that tell us? It tells us much about the great values that we have always cherished and lived by: tolerance, empathy, compassion, civility, hospitality, co-existence, intellectual partnerships between our cultures and other cultures, religious openness; we carry within us a powerful intellectual, aesthetic, and ethical heritage. Our young people need to learn about that.

6 MULTI-CULTURAL EDUCATION MAKES EVERYONE VISIBLE: NO-ONE IS INVISIBLE

The respect and integrity of every community, every culture, and every individual in this beautifully diverse society is crucial to nurturing pluralism and democracy. Dominant groups should not be invisible; heads of companies should not be invisible; multinational companies' plans in wonderful Bahrain should not be invisible. I want to be able to walk into a construction company, as an educator, and ask them: what are you doing to improve the quality of life in Bahrain? I have a feeling that I will have the backing of the government by the mere fact that this conference is taking place.

Everyone is visible and minority cultures and workers cannot be kept at the periphery, getting less than minimum wages and not even enjoying 1% of the benefits of their own work as they build high-rise after high-rise. We need to instil in the constitution of Bahrain laws that prevent racism and penalize racist practices severely against challenged poorer *ex*-patriate communities. Bahrain is endowed with rich and incredible diversity; its multicultural society and its ethnic communities constitute a beautiful mosaic. We already have a benevolent government that encourages diversity and the celebration of difference. Look at the rich cultural spring fests around you. Many music concerts and artistic exhibitions can be attended free of charge in Bahrain which is an unusual thing in this part of the world; it reminds me of Canada. This Spring Art & Culture Festival was envisioned by Shaikhah May Al Khalifa, yet another example of great contributions by women and another instance that demonstrates why we need to listen more carefully to women. Let us remember that it is in Bahrain that the first Gulf women worked and got an education. This is the right place for leadership, the right place for reform in higher education. This mosaic needs to be more and more inclusive, however.

7 CHILD-FRIENDLY BUDGETS AND PROTECTIVE ENVIRONMENTAL PROJECTS BY MULTI-NATIONAL COMPANIES: CREATING NON-GOVERNMENTAL ORGANIZATIONS THAT CAN HELP GOVERNMENT BY MENTORING, OFFERING CONSULTANCY SERVICES, AND MONITORING THE INFRACTIONS OF MULTI-NATIONAL GIANTS

The government needs to create laws that force companies to develop child-friendly budgets and action plans; my suggestion would be that no company should be allowed to work in Bahrain unless they are able to contribute something to this society, something that is visible, that is useful, and that is effective. The Ministry of Labour could, perhaps in collaboration with the Ministry of Education, create a competition whereby companies which can put together the most interesting and most creative projects for the protection of the environment in Bahrain could win certain prizes.

We need to put caps on profits made by huge multinational companies, companies that benefit from the laxity of the law with respect to *ex*-patriate workers from challenged and poor societies. Every company that benefits from being in Bahrain needs to give something back to Bahrain. This could be given in the form of some creative work in the area of environment protection: planting trees, cleaning up the beaches, creating green parks and amusement parks for children, giving "*zakat*" money to institutions that care for poorer and victimized sectors of society, for example: *The Be-Free Center*. The list is long. We need to put the idea of a child-friendly budget—a brilliant idea that I heard about in a conference launched by the United Nations last year here in Bahrain—into effect immediately. We also need to create more non-governmental organizations whose members would be local citizens to monitor these ongoing environmental projects and monitor the infractions of the multinational gigantic companies.

There is acute consciousness among our young of disparity in the social economic status among Bahrainis; there is also awareness of difference. The most important thing, however, is that there is a very fertile ground for the celebration of difference. At this juncture, we need to instil in our students the notion of celebrating difference and guarding human rights, concomitantly; otherwise, we will move backwards and not forwards.

8 THE QUALITY OF EDUCATION BAHRAINI UNIVERSITIES OFFER AND THE CRITERIA THEY FULFIL—RATHER THAN THE NUMBER OF UNIVERSITIES—DETERMINES WHAT IS GOING ON IN HIGHER EDUCATION

We see a plethora of universities mushrooming in Bahrain. They are quite competitive. Many are vying for the attention of the young with promises to teach them skills in the fastest ways possible. Packaged courses are sent from the West with promises to make a person an accountant in so many days, an IT manager in two weeks or in ten days, etc. Recently, a Bahraini theatre group did a play about such issues; it was presented at Al-Bareh Gallery and it was fantastic! I enjoyed it so much. We cannot globalize minds, this is a myth! And we certainly cannot become a global village in ways that are more focussed on commercial gain rather than on humanitarian values of inter-connectedness. The term Global Village becomes a farce if our globe is ruled by a prosperous few who own more than 80% of the resources of the world in which the rest of the world does not share!

When we open more-and-more universities without being able to find jobs for our young people, we are creating a situation of tension that will encourage our young to emigrate. When we treat certain communities of *ex*-patriates from the developing world as if they are less than human, then they will inevitably not have any kind of loyalty for the work nor for the country. The consequences could be grave.

Compartmentalized and packaged courses that have nothing to do with the language, culture, or heritage of local students are being imported into this country and all Gulf countries. A colleague of mine is writing a doctoral dissertation on the crucial importance of sensitizing teachers to the language, culture, and learning habits of students. Teachers from developed nations need to humble themselves, need to learn the language of the students, if not the spoken language, perhaps the traditions, cultural narratives, values and beliefs because language is embedded in culture. If students are not involved in the process, and if the habits, cultures, and attitudes of the teacher are foreign to them, they resist learning.

Packaged education only feeds the pockets of those institutions which oversee those programmes. Students do not, in the final analysis, benefit from packaged courses nor do they find the right jobs after finishing their studies. Many end up working in "call centres" instead of working in, for example, the Ministry of Environment, or the Ministry of Culture, instead of participating in an indigenous constructive programme that builds Bahrain on some level, instead of becoming independent business entrepreneurs owning small businesses. Graduates need encouragement and help in financing their endeavours because many have highly imaginative visions, and beautiful dreams and hopes. They need to feel that they are really taking part in building their country, Bahrain; construction of high-rise buildings is only one small way of building a society; building a society is all about building human resources, about building the whole person and the healthy nurtured and balanced person within society.

That is why we need an overall plan that is devised by both the public and private sectors together, a plan that the government will definitely seek to guard and protect. There is nothing worse than giving a student mere vocational training without a context instead of a good holistic engaging education in which she or he are participants and active learners, creators of their own futures. The student with good technique and good technological skills but no critical thinking may find out after emerging from graduate school that, the words of Sam Keen (1970), "She or he is empty of enthusiasm; that they have a profession but nothing to profess, that they have knowledge but no wisdom, ideas but few feelings, that she or he is rich in techniques but poor in convictions, the student may have an education but has lost an identity!"

If we do not protect our communities, our environment, our cultures, our values, and our heritage, we will not succeed in building a healthy future society. We have already begun paying the price, as many scholars at the 2007 "Conference on Arab Thought" kept iterating.

9 NURTURING AN UNDERSTANDING OF EMOTIONAL INTELLIGENCE: ENCOURAGING MULTIPLE PARADIGMS OF CREATIVE THOUGHT WITHIN DIVERSE CAREERS IN MULTIPLE FIELDS SINCE LEARNERS HAVE MULTIPLE INTELLIGENCES

We know from Howard Gardner's work in the book: *Frames of Mind: the Theory of Multiple Intelligences* (Gardner, 1983), that human beings do have multiple intelligences. We also know that we need "emotional intelligence" (Goleman, 1995) to recognize the diversity and multiplicity of human gifts and talents, to recognize the richness of the 'Other,' she or he who is not us and who is different from us, to celebrate and not denigrate the 'Other'; to acknowledge that unless we share in the diverse intelligences, cultures, narratives, values, traditions, and histories of the 'Other' we will remain limited, narrow-minded, exclusivist, and stultified; we cannot move forward in a constructive, dignified, and educative manner.

Everything I have discussed in this paper points to developing and nurturing our emotional intelligence in order to feel and understand with empathy, compassion, and with a comprehensive spirit the multiplicity of ways humans learn, the multiplicity of gifts humans possess, and the multiplicity of aspects through which we can participate in re-visiting and enriching our civilization as well as contributing to human civilization at large. We need to nurture in the minds and the hearts of students an education in *inter-connectedness* and *inter-dependence*, an education in *compassion* and a celebration of difference.

We also need to raise our male children to be sensitive and allow them to express their emotions when they need to, not oppress them with words like: "A man does not cry!" We need to allow our women to be stronger without calling them 'male wanna-bees' or accusing them of emulating men. We need to recognize the special gifts that each gender brings to the table and to education. We need to reverse the roles so that men can be nurturing fathers and women can be imaginative career makers. We have a lot of work to do in that domain.

We need to embed in our students' minds an image of the life as a circle and not a hierarchy; to empower our students with the knowledge that whatever kind of intelligence God has endowed them with, one is not better than the other: it is all a reflection of our humanity and richness. We need to get rid of the positivist paradigm's influence on education that has emphasized work in the sciences, in medicine and engineering, for example, and compromised the value of art, aesthetics, literature, poetry, and the humanities. We need to make our male students understand that there is a flawed perspective in their socialization that imposes on them behaviour that is devoid of feelings, that hides their compassion, that disables them from exhibiting their sensitivity, that labels them and stereotypes them if they are soft-hearted, if they are kind and gentle, if they show their empathy towards others. Females who are raising children also need to learn about that in order not to raise chauvinistic male children who become oppressors of women.

We need an education in compassion and in sensitivity that is linked to a constant conversation in universities and institutes of learning about ethics and moral values. We need to hold public debates about topics that have moral dilemmas in them, debates in schools and universities, on radio and television, where elders join the young in a healthy exchange, where teachers and parents are in constant conversation, where school and government administrators listen carefully to parents, teachers, and students.

Teaching civic duty and moral values is not an outmoded thing. We need to engage in ethical debates about diverse issues in order that all members of society continue to feel a sense of loyalty to this country no matter what their religion, race, gender, sect, culture, or socio-economic status. We need to give educators and educational institutions more voice in the media; to fund educational institutions, conferences, seminars, and to send more Bahrainis to participate in conferences around the world and to study as exchange students.

10 MAKE TECHNOLOGY SUCCUMB TO OUR DREAMS AND SERVE OUR ASPIRATIONS RATHER THAN CONTROL US

There must be within this audience visual artists, historians, journalists, mathematicians, teachers of literature, writers, poets, early childhood education specialists, physicists, chemists, business men and women, entrepreneurs, musicians, teachers of music, lab technicians, administrators, psychologists, psychiatrists, doctors, surgeons, etc. What distinguishes those who are imaginative from those who are not is the way they conceptualize, visualize, and think about their fields; imaginative thinkers find unusual ways to think about education within their fields, and those unusual ways that they come up with tend to be effective and give results in society.

We have come to this conference in an attempt to prod ourselves to be imaginative in the way we think about higher education. We have come together, administrators, teachers, professors, policy makers, government ministers, and specialists in different fields to see whether we can all enter into diverse dialogues in inter-connected ways in order to visualize in unusual and effective ways the future of higher education in the Gulf, but specifically in Bahrain. Thinking about a conference like this was highly imaginative on the part of the organizers and under the wise patronage of His Majesty, King Hamad Bin Isa Al Khalifa. Let us go forward together, hand in hand: let us have things to profess and not merely professions; let us be full of passion and have strong feelings about education and not only uphold hold ideas; let us be rich in convictions and not just rich in techniques; let us strive for wisdom and not only for knowledge.

REFERENCES

Ateek, N. 1989. *Justice and Only Justice: A Palestinian Theology of Liberation.* Maryknoll, NY, U.S.A.: Orbis Books.

Boullata, I. 1990. *Trends and Issues in Contemporary Arab Thought.* New York: State University of New York Press.

Campbell, J. 1973. *The Hero with a Thousand Faces.* Princeton, New Jersey: Princeton University Press.

Capra, F. 1991. *The Tao of Physics.* London: Flamingo.

Dewey, J. 1902/1990. *The School and Society; the Child and the Curriculum.* Chicago & London: The University of Chicago Press.

Dewey, J. 1938. *Experience and Education.* New York: Collier Books.

Dewey, J. 1934. *Art as Experience.* New York: Pedigree Books.

Egan, K. & Nadaner, D. (eds) 1988. *Imagination and Education.* Teachers College Press: New York.

Flowers, B.S. (ed.) 1988. *Joseph Campbell: The Power of Myth with Bill Moyers.* New York: Doubleday.

Freire, P. 1970. *Pedagogy of the Oppressed.* New York: Continuum.

Gardner, H. 1983. *Frames of Mind: The Theory of MultipleIntelligences.* New York: Basic Books.

Goleman, D. 1995. *Emotional Intelligence: Why It Can Matter More Than IQ*? New York: Bantam Books.

Hourani, A. 1991. *A History of the Arab Peoples.* New York: Warner Books.

Keen, S. 1970. *To a Dancing God.* New York: Harper & Row.

Maalouf, A. 1984. *The Crusades Through Arab Eyes.* London: Al Saqi Books.

Maguire, D.C. & Fargnoli, N. 1991. *On Moral Grounds: The Art/Science of Ethics.* New York: Crossroad.

Noddings, N. 1992. *The Challenge to Care.* New York: Teachers College Press.

Said, E. 1978. *Orientalism: Western Conceptions of the Orient.* New York: Vintage Books.

Said, E. 1993. *Culture and Imperialism.* New York: Vintage Books.

Said, E. 1994. *Representations of the Intellectual.* New York: Pantheon Books.

Said, E. 1997. *Covering Islam: How Media and the Experts Determine How We See the Rest of the World.* New York: Vintage Books.

Tarnas, R. 1991. *The Passion of the Western Mind: Understanding the Ideas That Have Shaped Our World-view.* New York: Ballantine Books.
http://www.globalpolicy.org/globaliz/define/2004/04heldinterview.htm Accessed on 27 May 2007.

BIBLIOGRAPHY

Christian-Smith, L.K. & Kellor, K.S. 1999. *Everyday knowledge and Uncommon Truths: Women of the Academy*. Oxford: Westview Press.
Clandinin, D.J. and Connelly, F.M. 1995. *Teachers' Professional Knowledge Landscapes.* New York: Teachers College Press.
Clandinin, D.J. and Connelly, F.M. 2000. *Narrative Inquiry: Experience and Story in Qualitative Research.* San Francisco: Jossey-Bass.
Dewey, J. 1887. My pedagogic creed. *The School Journal*, 54(3).
Dewey, J. 1916. *Democracy and Education: An Introduction to the Philosophy of Education.* New York: The Free Press. (Also 1916, Macmillan Company; 1944 John Dewey).
Gilligan, C. 1982. *In a Different Voice: Psychological Theory and Women's Moral Development.* Cambridge, Massachusettes: Harvard University Press.
Palmer, P. 1998. *The Courage toTeach.* San Francisco: Jossey-Bass.
Philips, C. 2001. *The Socrates Café: A Fresh Taste of Philosophy.* New York: W.W. Norton.
Ronald-Martin, J. 1994. *Changing the Educational Landscape.* New York: Routledge.

Higher Education in the Twenty-First Century: Issues and Challenges – Al-Hawaj, Elali & Twizell (eds)
© 2008 Taylor & Francis Group, London, ISBN 978-0-415-48000-0

University-based knowledge networks in the GCC: Opportunities and challenges

Abdelkader Daghfous & Noor Al-Nahas
School of Business and Management, American University of Sharjah, UAE

ABSTRACT: Collaborative knowledge networks come in various shapes and forms, digitally connected or clustered in a particular geographical location. In industry—university collaboration, a new model has emerged with more direct involvement and benefits for universities especially in Business Schools, focussing on managerial and administrative innovation. Yet, little research has been devoted to investigating the role of universities as sources and even co-ordinators of managerial innovations. The most popular among such innovations seem to be those related to IT-enabled enterprise management systems such as CRM (Customer Relationship Management) and SCM (Supply Chain Management) (Roberts, 2007). In SCM, various US universities have established forums and industry consortia with various companies and even government organizations. Such consortia explore new research avenues and bring the industry into the classroom, involving faculty and students in finding solutions to real industry problems of member companies. Since almost every university in the GCC has a Business School, the application of this new collaboration model in GCC universities is pertinent. This paper addresses the issues of transferability and implementation challenges of such university-based knowledge networks in the GCC.

1 INTRODUCTION

Knowledge-driven competition is imposing a pressure on firms to innovate at the product, process (Hauser et al., 2006), and strategic levels (Smith & Bagchi-Sen, 2006). Since extensive internal R&D is typically too costly, cheaper alternatives would include continuous learning programmes, greater involvement in external knowledge networks, and the development of corporate gatekeepers to reduce the communication and relative absorptive capacity gaps between the knowledge buyers and sellers (Lane & Lubatkin, 1998).

Industry, governments, and universities are joining hands in what is known as the triple-helix model in order to leverage disparate heterogeneous knowledge sources through knowledge generation, dissemination, and sharing to yield increased innovativeness and, thus, competitiveness (Carayannis & Alexander, 1999; Gilsing & Nooteboom, 2005). The growth of research in the areas of innovation and especially Knowledge Management has provided a fresh perspective on technology transfer and R&D management. The proliferation of e-business models has also generated new ideas and research on web-based knowledge exchange and the use of internet technology as a platform for collaborative innovation (Kostas et al., 2005).

Companies almost never innovate in isolation (Swink, 2006). Rather, their innovation activities are embedded in a network of different actors and "institutional" framework conditions. Therefore, it is not appropriate to consider only the innovation activities of companies. Instead, all activities in the entire innovation system starting from knowledge generation up to market introduction and penetration of new products, processes or services should be taken into consideration (Lundvall et al., 2002). The connect-and-develop networks employed by Procter & Gamble, wherein good ideas are fetched externally through collaboration with other businesses and individuals and brought in to capitalize on internal capabilities, is a recent example of this trend (Huston & Sakkab, 2006). The importance of universities stems from their traditional role as providers of both the scientific knowledge used as a basis of industry innovation through R&D and the intellectual or human capital that would ultimately carry out such innovation (Gunasekara, 2006; Mansfield, 1995).

Knowledge networks, defined as relationships based on knowledge, allow companies to reap the benefits of *potential cognitive synergies* (Canzano & Grimaldi, 2004). Members in the knowledge network are able to achieve significant economic value and other knowledge-based advantages resulting from the application of university-generated knowledge. Consequently, we have witnessed an increasing number of knowledge networks, though in a variety of forms, in almost every country as part of the required infrastructure for innovation, industrial development, and economic competitiveness (Scheel, 2002). There also seems to be a strong trend towards clustering in almost all industries as business entities increasingly agglomerate around universities and research centres (Cooke, 2001).

Recently, several American and European universities have been experimenting with new models of knowledge networks, wherein the university acts as a knowledge seller as well as a broker. For instance, Symbion, an award winning Science Park in Denmark, provides a flexible setting for coordination among start-up companies, the world of research, universities, and established businesses in the areas of ICT, bio-technology, and medical technology. At the opposite side of the spectrum, the Teradata Center for CRM at Duke University exemplifies how a single IT solution provider could partner with a leading business school to generate and disseminate CRM-related knowledge.

Table 1. Examples of knowledge networks in US Business Schools.

Consortium characteristics	NC state university supply chain resource consortium	Penn state center for supply chain research	The center for CRM at Duke university
University/School	The Supply Chain Management (SCM) programme at NC State University	Division of Research of the Smeal College of Business at Penn State	Located at the Fuqua school, funded by Teradata, NCR
Mission/Objectives	Applying the latest research, experience, and knowledge to identify executable solutions and capture practical learning in the process	Dedicated to research and education in the field of logistics	Fostering an appreciation of CRM Support world-class research on CRM Merge the best of theory and practice in CRM
Example of member companies	American Airlines, BP, Chevron, Shell	Unilever, Verisign, Whirlpool	NA
Benefits to companies	Get access to student projects Access to fresh talent Identify and tackle SC challenges	Access to research, symposiums, benchmarking data	Access to white papers, working papers, and conference presentations
No. of affiliated companies	15	46	1 (Teradata)
No. of faculty members	7	49	3
Example of events/Activities	Curriculum changes are being formulated to support the SCRC vision Research work on inventory process analysis & reducing inventory in VMI processes	Supply chain council Annual surveys Benchmarking consortiums	Workshop on Customer Base Analysis Tutorial on Applied Probability Models in Marketing Research: Introduction

This paper investigates the current and potential roles of local universities in GCC-based knowledge networks. Subsequently, we attempt to formulate policies and strategies that would help these universities develop knowledge selling and brokering capabilities in existing and future knowledge networks. More specifically, we focus on university-based consortia related to Supply Chain Management (SCM) since it is assuming an increasing importance in GCC cities, such as Dubai. This focus on SCM is consistent with the worldwide trend of competition among supply chains as opposed to individual companies (Yee & Platts, 2006). We address the questions of local universities could develop the capabilities necessary for hosting successful SCM consortia similar to those in MIT, NCST, Michigan State University, and Teradata Center for CRM at Duke University. Table 1 outlines the basic characteristics of a sample of such networks.

Related research about the Middle East and the GCC has been done primarily in the fields of economics and public policy. Most of such studies have been performed by agencies such the United Nations Development Programme (UNDP), UNIDO, Madar Research (based in Dubai), and graduate theses. Besides the wide range of academic and publicly funded studies produced in Western countries during the last two decades, there have been a number of quite sophisticated and rigorous studies produced by and about Latin American countries.

2 KNOWLEDGE NETWORKS AND INNOVATION

Innovations (product, process, organizational, or strategic) are increasingly involving a multitude of networks of knowledge-driven organizations. Such networks come in various shapes and forms, digitally connected or clustered in a particular geographical location. The extensive literature on industrial clusters, science and technology parks, knowledge marketplaces, communities of practice, and R&D consortia has provided significant evidence of the benefits of such knowledge networks (Chesbrough, 2003). These benefits range from commercial innovations to job creation, as well as the increase in the absorptive capacity of the recipient organizations. In addition, they help companies focus on their core competencies by "outsourcing", in a sense, the knowledge creation activities to an organized body of collaborating firms, of which it is a member itself (Quinn, 2000). Therefore, firms need to learn how to exploit internal and external knowledge effectively in order to develop the requisite absorptive capacity that enables them to successfully innovate and, as a result, develop a sustainable competitive advantage in the global marketplace (Daghfous, 2004).

Knowledge that is less tacit, more codifiable, and less path-dependent tends to be highly mobile in knowledge networks (Tallman et al., 2004). The challenge in knowledge networks is associated with the creation and exchange of tacit knowledge. The extent to which this knowledge is exploited is highly dependent on the absorptive capacity of the partnering firms, wherein absorptive capacity is defined as the ability not only to obtain knowledge, but also to combine it with existing knowledge, comprehend it, transform it, and use it (Zahra & George, 2002). Absorptive capacity of the various partners of knowledge networks determines the value that these partners can gain from the initiative. The different knowledge generation activities, in the various types of knowledge networks, not only provides new knowledge to participants but also makes these firms more capable of assimilating and using related knowledge and technologies (Cohen & Levinthal, 1990).

Recent approaches to analyse innovation systems emphasize the high relevance of strategic co-operation among different actors in innovation processes. In addition, the generation of knowledge and learning by individuals and organizations is regarded as a vital part of innovation systems as well. In contrast to codified knowledge, tacit knowledge cannot be easily transferred because it has not been codified in an explicit form. Since codification is never complete, some forms of tacit knowledge will continue to play an important role, in particular in high-technology fields. This tacit knowledge can only be transferred through face-to-face interaction and employee mobility. Thus, innovation systems literature has paid increasing attention to formal and informal co-operation and direct interaction among firms. In addition, the role of relationships between companies and non-firm organizations as a source of innovation has been emphasized in several studies.

3 COMMON FORMS OF KNOWLEDGE NETWORKS

Science parks, technology parks, and research parks are institutional mechanisms formed for the purpose of supporting a culture of innovation and competitiveness for its members. Their aim is to create operational science and technology-based links for companies with research centres and higher education institutions to facilitate the formation and growth of knowledge-based industries. They typically have an active management body that focusses on fostering the transfer of collaborative technology and research outcomes to organizations, the incubation of spin-off processes and ideas, and the provision of high-quality spaces and facilities for member institutions (Durao et al., 2003).

Meanwhile, industrial clusters are high-performance networks that link local companies to complementary industries, academia, venture capital, financing, funding, and the government. The goal is to enable local companies to learn best practices and provide the collective resources necessary for integration with World Class clusters and systems to acquire and share knowledge and technology, and gain access to new markets (Scheel, 2002). Such clusters are usually located in the proximity of universities and training and research centres to facilitate interaction and exchange between business entities and the sources of up-to-date knowledge and services, which become players in the industrial development process (ESCWA, 2006). The ultimate result is higher innovation and competitiveness, leading to better industrial development for the region in which the industry cluster exists.

In contrast, incubators are organizations that provide the necessary infrastructure as well as the management and operation processes necessary for pre-incubation and incubation of innovative entrepreneurial ideas up to the point of spin-out of technological enterprises. The services provided by these incubators also include support and sustaining activities, venture capital programmes, risk-sharing programmes with the financial sector, and promotion programmes (Scheel, 2002).

Communities of practice can be defined as "self-organized groups … generally initiated by employees [with complementary knowledge] who communicate with one another because they share common work practices, interests, or aims" (Davenport and Prusak, 2002). Such communities can support the codification and transfer of tacit knowledge through sharing ideas, exchanging experiences, and capturing lessons learnt. Participants in communities of practice develop a shared identity as they generate and exchange specialized knowledge, both tacit and explicit, from evolving shared practices (Tallman et al., 2004). Though mostly informal in nature, these groups may formalize through time, assigning themselves a title, and establishing regular routines of knowledge exchange (Neef, 2005).

Consortia are knowledge creating bodies that are either industry-based or university-based. There has been a growth in the number of research consortia over recent years in an attempt by firms to seek outside innovation and technologies (Roberts, 2007). Consortium members are typically not bound by legal contracts. Rather, they base their collaboration on trust and mutual interest. In fact, types and sizes of consortia abound, with some of them being called "networks" or "partnerships" instead. The ultimate goal is the achievement of a common goal that cannot be achieved efficiently without cross-organizational collaboration and sharing of resources such as skills, knowledge, and physical assets in an attempt to eliminate the duplication of efforts (Carayannis & Alexander, 1999).

4 UNIVERSITY AS A SOURCE OF INNOVATION IN MANAGEMENT PROCESSES

The existence of knowledge networks (or sometimes called spaces) is an advantage upon which a country or a region can build environments favourable for innovation in specific fields relevant to social and economic development. As Porter and Stern (2001) put it, "Innovation has become the defining challenge for global competitiveness. To manage it well, companies must harness the power of location in creating and commercializing new ideas." The ultimate goal of this study is to make the best use of available resources by maximizing the effectiveness and efficiency of local knowledge markets, thereby making this region a destination for innovating firms. Porter & Stern (2001) also added that "universities are perhaps the single most important institutions linking a nation's clusters and the common innovation infrastructure." Indeed, this study highlights the central role that universities could play in building the region's innovation capacity.

The literature on industry—university collaboration has thus far viewed universities as sources of new scientific and engineering inventions that ultimately lead to the introduction of new products and the development of new companies (Marri et al., 2002). Support for this form of knowledge production has been declining over the years, with the market demanding more professional education, knowledge production, and innovation in management and organizational practices (Horn, 2005). Little research effort has been devoted to investigating the role of universities as sources of innovations in the form of process and enterprise management systems such as CRM and SCM (Roberts, 2007).

However, it is often argued that universities can only excel in this domain in "narrowly defined niches" (see, for example, Huff, 1999). As indicated earlier, the focus in this research is on the role of universities in SCM-related knowledge networks in order to show how universities might possibly play a more significant and effective role in contributing to process management innovations. In fact, participation in knowledge networks is essential for universities to effectively contribute to such process innovation because theoretical organizational knowledge alone is of value only when augmented with hands-on experimentation with organizational tasks and direct connections with companies.

Universities provide access to new ideas, specific faculty, and students as prospective employees. In addition, universities provide companies from various industries with access to university-based research in their fields. Instead of providing task-specific knowledge and skills, universities can help participants develop skills for acquiring and processing knowledge in a knowledge-sharing environment so that employees are able to assimilate the needed knowledge quickly (Tu et al., 2006). In return, universities expect to receive funding to support faculty research, facilities, and graduate students. Also, universities look for internships and placement opportunities for students. More importantly, universities hope to gain access to proprietary data and specialized equipment.

5 WHY SHOULD A FIRM JOIN A KNOWLEDGE NETWORK?

Firms are engaged in collaborative R&D generally because it allows them to use external resources for their own purposes directly and systematically. The benefits of R&D co-operation (Robertson & Langlois, 1995) can be described as follows:

- Joint financing of R&D;
- Reducing uncertainty
  - Access to sources of new knowledge;
- Realizing cost-savings;
- Realizing economies of scale and scope.

The disadvantages of collaborative R&D are caused by transaction costs (Pisano, 1990 and Williamson, 1989) especially for co-ordinating, managing and controlling the activities of the different parties involved. Transaction costs are mainly related to the following topics:

- Co-ordinating distinct organizational routines, styles, etc.;
- Combining complementary assets, resources, etc.;
- Fixing transfer prices of intangible goods, for example information or know-how;
- Regulating the exploitation (appropriation) of the results (rates of return) of joint R&D.

In the context of university-based knowledge networks, potential benefits to member organizations include achieving the following in a cost-effective way, in addition to several other intangible benefits such as:

- First-hand opportunity to learn from a range of projects, activities, peers, and knowledge providers;

- Formation of win-win partnerships with other members to achieve a continuous stream of product, process, and strategic innovations;
- Involvement in leading-edge research projects with practical and commercial focus;
- Opportunity to test and experiment with innovative solutions in a real-world yet low-risk environment;
- Continuous access to a wealth of knowledge (experience, talent, skills, expertise, creativity) to gain a "first-mover" advantage in the market;
- Increase the absorptive capacity of individual participants as well as the organizations they represent;
- Gain a reputation of a cutting-edge and continuously improving organization in the industry.

6 KNOWLEDGE TRANSFER: NETWORK APPROACHES AND POTENTIAL BARRIERS

Praise et al. (2006) used Organizational Network Analysis (ONA) as a framework to characterize the *relational fabric* of organizations, as it is very important to consider in any knowledge-retention strategy. This relational fabric can be defined as the intertwined set of relationships that participants in a network hold towards each other and towards entities external to the network. The occurrence, nature, and intensity of these relationships depend on the roles played by participants in the network.

This study is consistent with Davenport and Prusak's (2002) earlier characterization of knowledge markets. Informal networks of practice are the best manifestation of knowledge market dynamics. They are relational webs through which people communicate, ask about knowledge sources, and get answers to their questions to get their work done. In knowledge markets, *brokers*, also known as gatekeepers or boundary spanners, are those who bring knowledge sellers and buyers together and facilitate such knowledge exchange. Knowledge *sellers* are those participants in knowledge markets who are known to have substantial knowledge and experience about certain subjects, and thus are sought constantly for information and insight. Meanwhile, knowledge *buyers* are those participants who seek knowledge that is necessary for accomplishing their work and improving their insights (Davenport & Prusak, 2002).

Effective knowledge transfer typically depends on three main factors (Daghfous & Kah, 2006): (1) willingness of knowledge buyer and seller to exchange knowledge; (2) existence and quality of transfer channels; and (3) absorptive capacity of the recipient. Knowledge buyers should be willing to acquire knowledge, and should perceive the knowledge source as credible and valuable. At the same time, knowledge sellers should have the ability and willingness to share their knowledge with others. Although availability of transfer channels is vital to achieving effective transfer, they are of no value if the recipient does nothing with the transmitted knowledge. It is, hence, beneficial and insightful to look at the adoption and implementation of knowledge networks, by GCC universities in the field of SCM and possibly CRM, as an institutional innovation. Absorptive capacity can be a facilitator or an inhibitor of such adoption and successful implementation.

6.1 *The absorptive capacity of organizations*

There have been several attempts to define the concept of absorptive capacity. Zahra and George (2002) defined absorptive capacity as a set of organizational routines and processes by which firms acquire, assimilate, transform, and exploit knowledge to produce a dynamic organizational capability. A second definition of absorptive capacity was presented as the capacity to learn and solve problems (Kim, 1997). A third attempt defined absorptive capacity as the firm's ability to identify, assimilate, and exploit outside knowledge (Yuko Kinoshita, 2000; Cohen & Levinthal, 1990). Despite the differences in defining absorptive capacity, all the above definitions include the act of interacting with the external environment in order to enhance an organization's competitive position.

6.2 *Elements of absorptive capacity*

There are four different but complementary dimensions of absorptive capacity, they are: (1) acquisition, (2) assimilation, (3) transformation, and (4) exploitation. It is important to note here that these four elements progress in chronological order, in that the first element needs to take place for the next one to ensue.

- **Acquisition** is the first component of absorptive capacity. It is defined as the ability of the organization to recognize, value, and acquire new external knowledge that is critical to a firm's operations (Zahra & George, 2002).
- **Assimilation** is the second component of absorptive capacity. It refers to the ability of a firm to assimilate new external knowledge. Assimilation can also be defined as a firm's routines and processes that allow it to analyse process, interpret and understand the information obtained from external sources (Zahra & George, 2002).
- **Transformation**, the third component of absorptive capacity, refers to thc firm's ability to develop and process the routines that facilitate combining existing knowledge with the newly acquired and assimilated knowledge.
- **Exploitation**, the fourth component of absorptive capacity, refers to a firm's ability to commercially apply new external knowledge to achieve organizational objectives (Lane & Lubatkin, 1998).

7 OPPORTUNITIES AND CHALLENGES IN THE *GCC*

This section begins with the role of GCC local governments, as key members in a broader university—industry—government coalition. This is also important since all major universities and research parks in this region are created and partially funded by the local governments. Table 2 shows the result of an exploratory investigation of the perception of some local service companies with respect to the expected benefits and concerns about their possible participation in an SCM or a CRM consortium. The companies in this preliminary study are essentially in service industries, which is a pillar of this region's future economic strategy.

7.1 *The role of the local government*

Today's business environment is epitomized by dynamism and high levels of competitiveness. The main impetus behind fostering the collaboration between universities and firms lies in its ability to render firms more competitive *via* its provision of the latest technological research (Marri et al., 2002). The understanding that regions with a larger stock of human capital produce a higher growth of knowledge and, consequently, a higher growth rate of output further supports the need for university and industry collaboration (Mathur, 1999). This phenomenon induces governments to take on new roles and responsibilities to incorporate a wider variety of functions, and tasks.

Consequently, governments are moving away from their traditional functions of financial providers, into their new roles of knowledge brokers and boundary spanners. An example of such new roles is some governments' efforts to develop policies or initiatives to minimize the obstacles that block the functioning of innovation systems (Solleiro, 2003). Hence, one can argue that evolving into a knowledge broker is a logical extension of governments' traditional roles and responsibilities. Other government roles include initiatives to prepare their entire population for participation in the knowledge economy, which necessitates tight collaboration with education initiatives (Igonor, 2002). Additionally, government support can also take the form of establishing colleges and research centres in industrial zones (Marri et al., 2002). The formation of advisory boards, consisting of government, industry, and university representatives at regional and national levels could also help build trust.

In this paper, we focus on the role of universities as central elements of knowledge networks in the Gulf Co-operation Council (GCC) region, where one of the primary concerns has been the high rate of unemployment of nationals. Another concern in this region is the creation of an attractive environment for Foreign Direct Investments (FDI) in an attempt to develop and nurture non-oil industries. To address these concerns and realizing the role of research in creating jobs, GCC countries have embarked on establishing knowledge networks of various forms to facilitate research and the transfer

Table 2. Expected benefits and concerns of GCC local companies about SCM and CRM knowledge networks.

Company	Expected benefits	Concerns
Hospital A	SCM focus: access to expensive and complex medical equipment, educate local doctors and minimize the need for patients to travel abroad for advanced medical treatment.	Quality of the consortium: excellence, mentality and attitude of its management, and how international it is.
Hospital B	SCM focus: access to more supply chain members (medical equipment suppliers, pharmaceuticals, catering, and IT/HIS companies) and the intensive experience of other hospitals and universities	Knowledge hoarding by other hospitals and supply chain members
Hospital C	SCM focus: attending seminars on supplier relationship management, cost optimization, e-procurement, and future logistics, and building skills and knowledge in materials management.	Lack of trust in the ability of such consortiums. Prefers to be a follower than a first mover in membership in a SCM consortium.
Hospital D	SCM focus: wish to do research with university students and other companies on cost optimization in the healthcare industry	Unless the consortium has clearly defined goals and outcomes. There's uncertainty associated with investing the time and money in such consortiums.
Hospital E	SCM focus: enhance personal contacts + learn international practices + awareness of other suppliers and customers + reduce bargaiing power of suppliers +	Level of consortium service + cost of participation + lead time enhancements and improved access to medical items + confidentiality of knowledge
Hotel A	SCM + CRM focus: interest in conducting research to facilitate supply chain choices and operations.	No clear and effective management for consortia + hotel contribution is limited + lack of specialization in the hotel industry
Hotel B	CRM focus: achieve best practices.	Unwillingness to share competitive and customer confidential data during experience exchange
Insurance A	SCM/CRM focus: identify and discuss CRM related problems and solve them in the context of the insurance industry	Difficult to measure the benefits of a consortium against the cost of investing in them
Insurance B	CRM focus: sharing CRM expertise and experiences related to the insurance industry	No measurable benefits
Computer A	SCM/CRM focus: gain fresh young knowledge + new research + ideas from young graduates + enhance public relations and social image	Location and facilities of the consortium
Computer B	SCM/CRM: interest in academic research.	CRM and SCM specialized consortiums are still too theoretical to happen in the region
Computer C	SCM/CRM focus: lowering the cost of R&D + supporting company image and public relations	Participation in consortiums will largely depend on the quality of other consortium members

of knowledge among businesses, governments, and universities. For instance, such existing networks include Dubiotech, Dubai Technology Park, and Dubai Silicon Oasis (DSO) in the UAE, the Qatar Science and Technology Park (QSTP) in Qatar, and the Oman Knowledge Oasis in Oman.

7.2 *The role of GCC universities*

Unlike their counterparts in the West, so far most of the GCC-based knowledge networks are rarely affiliated directly with major universities, with the exception of the QSTP (Qatar Science and Technology Park). At the QSTP, technology development in any field is welcome, although the focus is on those technologies that are most relevant to Qatar's economy and needs, namely gas and petro-chemicals, healthcare, information and communication technologies, water technologies, the environment, and aircraft operations.

A university-based knowledge network in the GCC may adopt one of several business models. Such a network would, to a certain extent, act as a knowledge seller, thereby competing with other existing knowledge sellers in the market such as:

- Existing consulting firms;
- Existing consortia such as the Supply Chain Logistics Group (SCLG) of the Middle East, which is a non-profit organization established in Dubai;
- IIR (Institute of International Research) Middle East based in Dubai, which organizes industry-focussed conferences, seminars, specialized training programmes, and industry exhibitions;
- GOIC (Gulf Organization of Industrial Consulting), based in Doha, Qatar, which provides client-focussed research, information, and consulting services for industrial companies such as in chemicals, metals, petro-chemicals, engineering, and food sectors;
- Sheikh Mohamed Bin Rachid Technology Park in Dubai, which is designed to be a regional technology and science hub, housing industry clusters in fields such as oil and gas, biotechnology, pharmaceuticals, etc.;
- Qatar Foundation's Science & Technology Park (QSTP), the main activity of which involves the development or transfer of technology. The QSTP includes firms like EADS, ExxonMobil, Microsoft, Shell, and Total.

However, this study focusses on a particular knowledge domain, namely Supply Chain Management (SCM). During the past two decades, companies in this region have made significant investments in ICT. There is equally significant body of literature that examined IT and Information Systems (IS) adoption and implementation from various perspectives (e.g., Zmud, 1982, 1984). Most of these studies, however, have focussed on technology adoption and implementation. In comparison, little attention has been paid to the adoption and implementation of IT-enabled organizational innovations, such as Total Quality Management—TQM—(e.g., Ravichandran, 2000), Customer Relationship Management—CRM—(e.g., Raisinghani, et al., 2005), and Supply Chain Management—SCM—(e.g., Sengupta et al., 2006). These intertwined areas are usually examined separately, although they converge towards service improvement and customer satisfaction.

Supply Chain Management spans all movement and storage of raw materials, work-in-process inventory, and finished goods. It focusses on managing the flow of services and information through the supply chain in order to attain the level of synchronization and collaboration that would make it more responsive to customer needs, while lowering total costs. Companies that have excelled in the implementation of SCM processes, analytical tools, and technologies have enjoyed several benefits such as:

- Greater control over suppliers and their quality standards due to strong relationships, better co-ordination and collaboration among supply chain members;
- Better streamlined processes, shorter lead times and replenishment, and enhanced equipment readiness and utilization;
- Enhanced communication and co-operation among members of the supply chain, leading to enhanced product/process designs;

- Faster customer response and improved delivery performance;
- Improved forecasting accuracy, planning and scheduling capabilities;
- Higher productivity and better responsiveness to demand fluctuations;
- Lower levels of inventory throughout the chain and substantial cost reduction.

We consider universities as active members in domain-specific knowledge networks, which essentially are strategic communities of practice leading to organizational innovation in the area of SCM (Storck & Hill, 2000). The question, then, is whether universities in the GCC are currently capable of playing the role of knowledge sellers and/or brokers in SCM focussed knowledge networks, wherein the buyers would be organizations located in the region. Here, we rely on the absorptive capacity perspective presented in section 6, where external knowledge is essentially knowledge about the value and the process of establishing a university-based SCM knowledge network.

Based on the four components of absorptive capacity, we argue that local universities would need to have (i.e., develop) the ability to recognize the value of SCM and the value of knowledge networks to the university, the various concerned industries, and community at large. This ability requires that local universities already have prior knowledge and R&D activities related to these knowledge domains, namely SCM and Knowledge Networks. The second component, assimilation, indicates that local universities have the requisite academic talent to analyse, process, and understand the complexities related to the institutional innovation in question. The third component, transformation, requires that local universities develop the ability to combine and integrate SCM knowledge networks with its existing mode of operations and curriculum. This may include adapting the business models found in knowledge networks based in Research-A universities to the local context, especially in terms of important constraints such as students' academic levels, faculty salaries, labour laws, and the absorptive capacities of local companies.

Finally, exploitation would require the alignment of the systems of faculty recruiting, evaluation, reward, as well as the teaching load distribution, with the new realities and demands of such a knowledge network. Taking the lead in SCM among many others, such as CRM, would enhance the image of the university as the most innovative and responsive one in the region. Creating and managing knowledge and local knowledge networks, however, involves a paradigm shift. The local university would have to change the way it manages its own intellectual capital. For example, it could start by changing the way it conceives community service, from inwardly focussed service to outwardly focussed formal and informal communities of practice.

The above analysis seems to converge towards a single critical factor that revolves around "talent". This also relates directly to the process of recruiting of faculty, higher administration and even support staff. As emphasized in the "absorptive capacity" literature, R&D is an essential ingredient in building an organization's ability to understand, assimilate and use new knowledge. Local universities should invest in such ability through recruiting faculty with quality research records and potential; and provide such faculty with the necessary resources, rewards, and administrative support. Our analysis also indicates that local universities would be ill-advised to ignore the need for aligning their organizational designs and processes with new innovations and strategic initiatives such as the establishment of knowledge networks. This alignment process is, however, in itself a learning process that also requires time and top management commitment.

8 CONCLUSIONS

The GCC is undergoing what appears to be a planned transition towards a knowledge economy. A number of industrial clusters, R&D initiatives, and educational institutions have already been established, while others are still in conception. This trend has proliferated throughout the GCC region. Therefore, it would be beneficial to explore how the triangular co-operation among universities, governments, and the private sector would enhance the creation, diffusion, and successful implementation of new knowledge in GCC countries and in countries where most companies lack the resources to embark alone on the innovation path.

This study was primarily motivated by a vision of the potential role that GCC universities can and perhaps should play to enhance the local human capital and become a central location for knowledge exchange and innovation. Indeed, we believe that several private as well as public GCC universities, with the appropriate public support, could act as liaisons between state-of-the-art knowledge and local firms, in order to bridge the absorptive capacity gap and promote various types of innovations. This exploratory research can be considered as a first step in a sequence of future inter-disciplinary studies aimed at developing insight, empirical evidence, and recommendations for more efficient and effective knowledge markets in this region.

Local GCC universities that may host an SCM knowledge network seem to have relatively modest research capabilities and resources in SCM and other business and management areas. Consequently, we argue that the research and knowledge generation and dissemination potential of local GCC universities is so far under-exploited, and that the benefits of such potential to the advancement of the socio-economic conditions and the creation of jobs are not fully realized. With the exception of a few research centres in Engineering and ICT schools, the notion of knowledge networks has been slowly adopted in the GCC, though without an adequate research and collaboration infrastructure and culture necessary to establish real knowledge networks. Nevertheless, the transfer of such institutional innovation is a trial-and-error process that requires universities to become learning organizations themselves, in addition to being institutions of higher learning.

REFERENCES

Canzano, D. & Grimaldi, M. 2004. Knowledge Management and Collaborations: Knowledge Strategy and Processes in the Knowledge Networks. *Proceedings of I-KNOW '04, Graz, Austria, June 30–July 2.*

Carayannis, E. & Alexander, J. 1999. Winning by co-opting in strategic government-university-industry R&D partnerships: The power of complex, dynamic knowledge networks. *Journal of Technology Transfer*, 24(2–3): 197–210.

Chesbrough, W.H. 2003. The era of open innovation. *Sloan Management Review*. Spring 2003. p. 35.

Cohen, W. & Levinthal, D. 1990. Absorptive capacity: A new perspective on learning and innovation. *Administrative Science Quarterly*, 35: 128–152.

Cooke, P. 2001. Regional innovation systems, clusters and the knowledge economy. *Industrial and Corporate Change*, 10(4): 945–974.

Daghfous, A. & Kah, M. 2006. Knowledge management implementation in SME's: A framework and a case illustration. *Journal of Information and Knowledge Management*, 5(2): 107–115.

Daghfous, A. 2004. Absorptive capacity and the implementation of knowledge-intensive best practices. *S.A.M. Advanced Management Journal*, 69(2): 21–28.

Davenport, T. & Prusak, L. 2002. *Working Knowledge: How Organizations Manage What They Know*. Boston, Massachusetts: Harvard Business School Press.

Durao, D., Sarmento, M., Varela, V. & Maltez, L. 2003. *Virtual and Real-Estate Science and Technology Parks: A Case Study of Taguspark*. Technical University of Lisbon.

ESCWA. 2006. Network of Technology Parks and Incubation Schemes: Definitions. Available at http://www.escwa.org.lb/ntpi/definitions.asp

Gilsing, V. & Nooteboom, B. 2005. Density and strength of ties in innovation networks: An analysis of multimedia and biotechnology. *European Management Review*, 2: 179–197.

Gunasekara, C. 2006. Reframing the role of universities in the development of regional innovation systems. *Journal of Technology Transfer*, 31: 101–113.

Hauser, J., Tellis, G. & Griffen, A. 2006. Research on innovation: A review and agenda for marketing science. *Marketing Science*, 25(6): 687–711.

Horn, P. 2005. The changing nature of innovation. *Research Technology Management*, 48(6): 28–34.

Huff, A. 1999. Changes in organizational knowledge production. *Academy of Management Review*, 25(2): 288–293.

Huston, L. & Sakkab, N. 2006. Connect and develop inside Procter & Gamble's New Model for Innovation. *Harvard Business Review,* March 2006.

Igonor, A. 2002. Success factors for development of knowledge management in e-learning in Gulf Region Institutions. *Journal of Knowledge Management Practice* (June).

Kostas, K., Mentzas, G., Apostolou, D. & Georgolios, P. 2005. Knowledge marketplaces: Strategic issues and business models. *Journal of Knowledge Management*, 8(1): 130–146.
Kim, L. 1997. *Imitation to Innovation: The Dynamics of Korea's Technological Learning*. Boston, MA: Harvard Business School Press.
Kinoshita, Y. 2000. R&D and technology spillovers via FDI: Innovation and absorptive capacity, *William Davidson Institute Working Papers Series* 349, William Davidson Institute at the University of Michigan Business School.
Lane, P., & Lubatkin, M. 1998. Relative absorptive capacity and inter-organizational learning. *Strategic Management Journal*, 19: 461–477.
Lundvall, B.-Å., Johnson, B., Andersen, E.S. & Dalum, B. 2002. National systems of production, innovation and competence building. *Research Policy*, 2: 213–231.
Mansfield, E. 1995. Academic research underlying industrial innovations: Sources, characteristics and financing. *The Review of Economics and Statistics*, 77(1): 55–65.
Marri, H., Gunasekaran, A., Kobu, B. & Grieve, R. 2002. Government-industry-university collaboration on the successful implementation of CIM in SMEs: an empirical analysis. *Logistics Information Management*, 15(2): 105–114.
Mathur, V.K. 1999. Human capital-based strategy for regional economic development. *Economic Development Quarterly*, 13(3): 203–216.
Ming Yu, C. 2002. Socializing knowledge management: The influence of the opinion leader. *Journal of Knowledge Management Practice* (December).
Neef, D. 2005. Managing corporate risk through better knowledge management. *The Learning Organization*, 12(2): 112–124.
Porter, M.E. & Stern, S. 2001. Innovation: Location matters. *MIT Sloan Management Review*, Summer: 28–36.
Quinn, J. 2000. Outsourcing innovation: The new engine of growth. *Sloan Management Review*, 41(4): 13–28.
Raisinghani, M.S., Tan, E., Untama, J.A., Weiershaus, H., et al. 2005. CRM systems in German hospitals: Illustrations of issues and trends. *Journal of Cases on Information Technology*, 7(4): 1–26.
Ravichandran, T. 2000. Swiftness and intensity of administrative innovation adoption: An empirical study of TQM in information systems. *Decision Sciences*, 31(3): 691–724.
Roberts, E. 2007. Managing invention and innovation. *Research Technology Management*, 50(1): 35–55.
Robertson, P.L. & Langlois, R.N. 1995. Innovation, networks and vertical integration. *Research Policy*, 24(4): 543–562.
Scheel, G. 2002. Knowledge clusters of technological innovation systems. *Journal of Knowledge Manamgenet*, 6(4): 356–367.
Sengupta, K., Heiser, D.R. & Cook, L.S. 2006. Manufacturing and service supply chain performance: A comparative analysis. *Journal of Supply Chain Management*, 42(4): 4–15.
Smith, H. & Bagchi-Sen, S. 2006. University-industry interactions: The case of the UK Biotech Industry. *Industry and Innovation*, 13(4): 371–293.
Solleiro, J.L. 2003. Fostering innovation and entrepreneurship. International experiences. *National University of Mexico*.
Storck, J. & Hill, P.A. 2000. Knowledge Diffusion through "Strategic Communities". *Sloan Management Review*, 41(2): 63–74.
Swink, M. 2006. Building collaborative innovation capability. *Research Technology Management*, 49(2): 37–48.
Szulanski, G. 1996. Exploring internal stickiness: Impediments to the transfer of best practice within the firm. *Strategic Management Journal*, 17: 27–43.
Tallman, S., Jenkins, M., Henry, N. & Pinch, S. 2004. Knowledge, clusters and competitive advantage. *Academy of Management Review*, 29(2): 258–271.
Tu, Q., Vonderembse, M., Ragu-Nathan, T. & Sharkey, T. 2006. Absorptive capacity: Enhancing the assimilation of time-based manufacturing practices. *Journal of Operations Management*, 24: 692–710.
Yee, C. & Platts, K. 2006. A framework and tool for supply network strategy operational-ization. *International Journal of Production Economics*, 104(1): 230–248.
Zahra, S. & George, G. 2002. Absorptive capacity: A review, re-conceptualization and Extension. *Academy of Management Review*, 27: 185–203.
Zmud, R.W. 1982. Diffusion of modern software practices: Influence of centralization and formalization. *Management Science*, 28(12): 1421–1431.
Zmud, R.W. 1984. An examination of push-pull theory applied to process innovation in knowledge work. *Management Science*, 30(6): 727–738.

Higher Education in the Twenty-First Century: Issues and Challenges – Al-Hawaj, Elali & Twizell (eds)
© 2008 Taylor & Francis Group, London, ISBN 978-0-415-48000-0

Enabling university excellence through a human resource strategy

Jim Horn
Director, Human Resources, Prince Mohammad Bin Fahd University, Al Khobar, Saudi Arabia

A key characteristic of our current global environment is the rapid rate of change driven primarily by advancements in technology. In this new world, sits the University. Are universities adapting and changing? Some people would say yes and would provide a few examples. Some people would hold the view that universities must remain true to their roots and not change. While others will say that if universities do not change, they will simply cease to exist. This paper is directed at the latter case. Universities need to change and change soon.

For the most part, life in a university has been a comfortable and quiet world. From time to time, there have been isolated incidents. Yet, if one looked carefully at how the University operates today and how it operated several years ago, one might make the case that little has changed. For the most part, faculty members are still using the same old traditional lecture methodology. And what about the curriculum—is it current and relevant? Is there a need for change? Who will make this decision—universities or society at large?

It is likely that the changes in a university will be driven by the same factors that have driven change in the private sector. (As an aside, the writer is conscious that the mere mention of the private sector—in the same breath as the University—creates an uneasiness in the academic community.) This notwithstanding, the point is that change in the private sector is primarily driven by competition. It is suggested that competition will also drive change in universities. The cost of tuition has risen dramatically and potential students have many alternatives to obtain employable job skills. Granted, a university serves many roles and is not simply a supplier for the job market. This is, however, one of the key expectations of employers, students and parents. If universities get too far out of touch with these realities, the student clientele will soon move elsewhere. Education is now big business and the private sector has entered in an earnest way. Tax payers in public institutions are questioning where the money is being spent. And a lot of money is being spent! Public accountability has been, and continues to be, on the rise. What is my son or daughter gaining by attending this particular University? Is your University prepared and capable of answering these questions? Can your University demonstrate quality and excellence? The public wants answers—meaningful answers—to these questions.

The university as an organization is the most complex organization in society. The cultures and sub-cultures are many. At the highest level, there are really two halves to the house. The academic culture and the professional or support culture. In many universities, these two cultures are divided and in conflict. This is especially obvious when resources are tight. If a university is to succeed and strive for excellence—as one community—then these inner divisions need to be addressed.

The decision making processes in a university (and for that matter in any organization) are not designed to handle system-wide change. The processes are designed to handle small incremental changes. Even if the President of the University was a visionary, he or she would have real difficulty in getting the University community to change. A prime example of this problem is the traditional experiences with strategic planning. The output and results were typically a new vision, mission and goals that were beautifully written, nicely framed and mounted on the University walls, and then simply collected dust. People who had to deliver these ideas were not involved, not engaged, and therefore did not support the change. The key is the process not the product.

How do you affect system wide change? How can you get people to support and sustain system-wide change? How can you get people to see the need for change and then identify what needs to be changed?

An integrated and progressive human resource strategy is needed in universities. A sound knowledge of organizational behaviour, organizational development, the learning organization and knowledge capital within the university leadership would be a good starting point. Awareness is the first stage. Applying the knowledge to the university as an organization is the next step. Application means the use of new process tools. A prime example is the relatively new field called "large group interventions." In other organizations, the leadership or catalyst for this responsibility is normally housed in the Office/Department of Human Resources. One of the tell-tale signs in any organizational structure is whether there is a Vice-President, Human Resources. This position is symbolic of the value that the organization places on its human resources. If you look at modern, successful organizations you will find that they have a very advanced human resources function. In the 1950's, you would find the "Personnel Office" reporting to the Head of the Finance Department. If an organization has this model today, they will surely be in trouble.

Universities need this same progressive model. There is a need for a senior human resources position reporting directly to the President. The objective would be to provide what one might describe as an integrated human resources model. This Officer would be responsible for providing services to both halves of the house in all areas of human resource management and development. Attached is one example of such a structure to illustrate the breadth and scope of an integrated function. The highest level and most strategic is the organizational development area. This area is designed to assist all levels of the University in change management processes. A human resource function is like a big spider-web. You touch one strand and it vibrates through the whole web.

Much has happened in our global environment in the area of change management over the past thirty years. Little of this knowledge and experience has been adapted and applied in universities. It is now timely for universities to benefit.

Universities claim that they are unique. And they are. Yet, there is a reluctance and fear of using some of the ideas and processes from the private sector that have worked. The fear is that the community of scholars will be destroyed. There is fear that the university will lose its role in society. The writer believes that these fears are unfounded and that the universities can change and remain unchanged. It is not an either/or proposition it is really a both/and. Some Universities are trying new approaches; not many, but some. Perhaps your University will be the next one?

Higher Education in the Twenty-First Century: Issues and Challenges – Al-Hawaj, Elali & Twizell (eds)
© 2008 Taylor & Francis Group, London, ISBN 978-0-415-48000-0

Quality of education in developing countries

Nabil Moussa
Ahlia University, Kingdom of Bahrain

1 INTRODUCTION

In this paper some of the essential factors for achieving high quality education in developing countries are studied. The concentration in this work is on the two most important factors, namely stressing critical thinking and enhancing the capability of using modern technologies in the education process.

The quality of education is analysed, not only at universities but also in schools in some developing countries. It is extremely important to start considering these two factors at school level since memorizing is widespread in many developing countries. Several capable instructors at universities in developing countries are handicapped in their teaching due to the fact that many students are lacking critical thinking and optimal usage of modern technologies.

In section 2 the current situation of education in some developing countries in terms of five important factors is analysed. In section 3 two proposals concerning the two main factors is given. Future aspects in the education process in developing countries are discussed briefly in section 4. Some recommendations are given in section 5 and finally, in section 6, the conclusion is stated.

2 CURRENT SITUATION

Although several developing countries are trying to raise the quality of education in their schools and universities, it is extremely important that more concentrated efforts be adopted in several directions.

Five of the most important factors are considered in the following:

2.1 *Co-ordination between schools and universities*

A close co-ordination between educators at schools and universities is lacking in several developing countries.

2.2 *Teaching methodology*

Memorizing instead of understanding is widespread at universities and especially at schools in many developing countries. This has an extremely bad influence on the progress of the students during their learning and studying, as well as in their careers (see Moussa 1991, 1995, 2004 and Krantz, 1993). Also the teaching in the classroom is very often in one direction, namely the teachers and instructors are talking and the students are listening and copying from the board. The very important "team work" is quite neglected as well.

2.3 *Usage of modern technologies*

This aspect is very poor in several developing countries. It is mainly due to lack of financial and technological resources as well as qualified teachers and instructors. Developing countries could make use of the experience of some developed countries where Information and Communication Technology (ICT) was intensively used in all levels of education especially during the last two decades (see Kent County Council, 2005; UCLA Center for Digital Innovation, 2005; Tatanen & Homaki, 2004; and Winer & Cooperstock, 2004). Under the patronage of His Majesty King Hamad bin Isa Al-Khalifa an official accomplishment of future schools in 11 public secondary schools in the Kingdom of Bahrain was announced on 18 January 2005 (see Bahrain News Agency, 2005, and an undated report by Dana & Hanadi).

2.4 *Curriculum development*

In the past, the quantity of material taught had higher priority over quality in several curricula at universities and especially at schools in many developing countries. The students were bombarded with huge amounts of material in almost all subjects and were consequently forced to memorize instead to understand. Recently some developing countries updated their curricula according to their needs (see Abdulkafi Albirini, 2004, and Pennington & Matthew, 1999) and some others made use of existing curricula in some developed countries (see Curriculum Corporation Technology, 2004). In the Kingdom of Bahrain efforts have been made to develop the schools' curricula (see Office of Education of the GCC, 1998).

2.5 *Assurance of high quality education*

There are efforts to assure high quality education in some developing countries. A good example is the project of "The regional UNESCO office for higher education in the Arabic Countries" to assure high quality education at universities in the Arab world (see Regional UNESCO Office for Education in the Arabic Countries, 2007). Also local efforts to assure high quality education in schools in some Arabic countries are to be considered.

3 TWO MAIN FACTORS FOR ACHIEVING HIGH QUALITY EDUCATION

The concentration in this paper is on the two main factors (given in sub-sections 2.2 and 2.3 above). The following sub-sections give a proposal concerning these two factors:

3.1 *Proposed teaching methodology*

Instructors at universities and more especially teachers at schools in developing countries should do their best to make their students minimize memorizing. Instead, teachers and instructors should make their students think critically, analyse logically, criticize objectively and propose creatively.

The developing countries should ensure that their teachers and instructors are capable of applying this proposed methodology in all schools and universities.

Lessons and lectures should be held in an interactive way with the students sharing their thoughts and opinions in the classroom. Also the teachers and instructors should let the students practise "team work" from an early age until graduation. The benefits to students, teachers, and instructors (especially females) will be:

1. Recognize their ability to think critically and discover how much better this approach is in comparison to memorizing;
2. Learn to share actively in interactively-conducted classes;
3. Practice working in teams from an early age;
4. Be helped to do homework and assignments independently and more effectively;
5. To become not blindly believing everything said by the teachers and instructors or written in books, but will start questioning till they become convinced;

6. Become able to propose new methods and/or solutions to certain problems in their study and in real life outside the classroom;
7. Become more creative in different fields of knowledge;
8. Be helped to become active citizens in their societies, not only during their study but also after graduation.

3.2 *Optimal usage of modern technologies*

Students, teachers and instructors should be guided and trained to become capable of optimally using modern technologies inside and outside the classroom. This necessitates the appointment of highly qualified experts to do the guiding and the training. Also it is necessary to equip schools and universities with up-to-date facilities of software, hardware and the Internet. The benefits to students, teachers and instructors (especially females) will be to

1. Be able to explore new resources, methods and techniques in their fields;
2. Communicate with other students, teachers and instructors at the national and international levels;
3. Gain essential skills to be able to deal with modern technologies;
4. Have social and economic advantages regionally while keeping traditions, and globally while being open minded;
5. Become more creative in different fields of knowledge using their gained ICT skills;
6. Be able to adapt to globalization and become active in the new world system.

4 NEW ASPECTS

Knowledge and ICT skills are at present quite important. In the future they will become crucial for the survival of people in developing countries. As globalization has a strong impact on the whole world and especially on developing countries, it is vital that all government and non-government institutions in developing countries study carefully all related issues to make use of the existing advantages and remedy some of the disadvantages of this new world system.

Consequently, achieving high quality education is essential and will be even more important in the future, especially the two main factors, namely critical thinking and optimal usage of modern technologies which will, in future, play an enormous role and must be intensively applied and developed by all the institutions concerned. This will become a necessity for survival of the people of the developing countries.

5 RECOMMENDATIONS

Some recommendations with different degrees of importance are as follows:

1. Establishment of a very close co-ordination between educators at school level and university level in developing countries is very highly recommended to assure homogeneity and application of similar goals, approaches and methodology;
2. It is highly recommended that developing countries adopt an approach of teaching their students at universities, and especially in schools, which stresses critical thinking, logical analysing, objective criticizing and creative proposing.
3. It is also highly recommended that developing countries build capacity of their teachers and instructors to become capable of teaching according to the proposed approach.
4. It is strongly recommended that students, teachers and instructors be guided and trained to become capable of optimally using modern technologies inside and outside the classroom. This necessitates the employment of highly qualified experts to do the guiding and the training;

5. The purchase of up-to-date facilities of software, hardware and networking are strongly recommended, to be made accessible to all students, teachers, and instructors;
6. Holding classes in an interactive way and letting students share their thoughts and opinions is also recommended;
7. Practicing "team work" from early an age is another recommendation;
8. Although some developing countries are revising and updating their schools' curricula, it is recommended that much more effort be undertaken towards updating the curricula taking into consideration the adoption of teaching the students to think critically and optimally use modern technologies;
9. Quality of the material taught must have higher priority than its quantity;
10. Developing a system for assurance of high quality education. This system has to be applied in developing countries either regionally or globally. It must have internationally-compatible measures;
11. Developing countries should deal with globalization in a positive way by adopting its advantages and remedying some of its disadvantages.

6 CONCLUSION

Developing countries should make intensive efforts to achieve high quality education at schools and universities in order to be able to survive the new world system, focussing on the two main factors discussed in this paper.

REFERENCES

Abdulkafi, A. 2004. Teachers' attitudes towards information and communication technologies: The case of Syrian EFL teachers. *Journal of Computers and Education* (Online).

Bahrain News Agency, 2005. HM the King inaugurates future schools. http://english.bna.bh

Curriculum Corporation Technology, 1994. *Report*. A Curriculum Profile for Australian Schools http://www.accc.edu.au

Dana, B. & Hanadi, Y. "A Proposal for Smart Schools in the Kingdom of Bahrain", *M.Sc. Project*, Ahlia University, Bahrain.

Kent County Council. 2005. *Report*. What is ICT? http://www.kentted.org.uk

Krantz, S. 1993. *How to Teach Mathematics: A Personal Prospective*. Providence, RI: AMS.

Moussa, N. 1991. An pproach for teaching discrete mathematics to computer science majors. *AMSE Review*, 17(1): 1–4.

Moussa, N. 1995. Teaching mathematical analysis to engineering freshmen and sophomores. *Proceeding of the Symposium on Science and Engineering Education in the 21st Century*, 202–209.

Moussa, N. 2004. Theoretical and Applied Approaches for Teaching Mathematics to Engineering Students. *WSEAS Transactions on Advanced in Engineering Education*, 1(1): 6–10.

Office of Education of the GCC, 1998. *Mathematics Books for Schools in the Kingdom of Bahrain* (Revision).

Pennigton and Matthew, 1999. *Asia Takes a Crash Course in Educational Reform*, Academic Research Premier. http://searcg.cpnct.com

Ratanen, P. & Ilomaki, L. 2004. Intensive use of ICT in school: Developing difference in students' ICT expertise. *Journal of Computers and Education*. http://www.sciencedirect.com

Regional UNESCO Office for Education in the Arabic Countries, 2007. "Questionnaire for a feasibility study for assurance of high quality education in the universities in the Arabic Countries".

UCLA Center for Digital Innovation, 2005. http://www.cdi.ucla.edu

Winer, L. & Cooperstock, J. 2004. The {Intelligent Classroom}: Teaching and learning with an evolving technological environment. *Journal of Computers and Education*. http://www.sciencedirect.com

Higher Education in the Twenty-First Century: Issues and Challenges – Al-Hawaj, Elali & Twizell (eds)
© 2008 Taylor & Francis Group, London, ISBN 978-0-415-48000-0

Reference framework for active learning in higher education

Pranav Naithani
BIT International Centre, Kingdom of Bahrain

ABSTRACT: The work presented in this paper traces the history of active learning and further utilizes the available literature to define meaning and importance of active learning in higher education. The study highlights common practical problems faced by students and instructors in implementing active learning in higher education and further identifies a set of individual practices being used worldwide to overcome the obstacles. Expectations and responsibilities of students and instructors are also specified to enhance the efficiency of active learning environment. The paper also traces the importance of student and instructor rapport for a successful learning environment.

Some of the important learning methods such as Problem Based Learning, Co-operative Learning, Assignment Based Learning are compared and analysed with an aim to inculcate and develop creative and analytical abilities amongst the higher education students. Students' active learning is linked to multiple intelligence and tools being used to exploit multiple intelligence are also identified.

The paper concludes with a list of suggested tools and pre-requisites for a successful active learning environment in higher education with specific reference to the importance of the role of the instructor.

1 INTRODUCTION

Can a person expect to get a car driving licence just by attending the theoretical classes? Let us hypothetically assume the person manages to get a licence, but will he be able to drive the car? No! To be a perfect car driver a person has to and does take real life driving classes. He develops new skills by actually driving along with the trainer.

Higher education aims at an optimal blend of knowledge and skills. It provides a licence to the student for driving successfully through the maze of what, when, why, who and how. Usually the next step after higher education is a professional career. This is why the role and importance of active learning, and not passive learning (in which an instructor depends solely or predominantly on passive lectures), is crucial throughout higher education.

2 ACTIVE LEARNING

Roots of active learning can be traced to Confucius (551–479 BC), who stated, "*I hear and I forget. I see and I remember. I do and I understand*". Socrates (470–399 BC) had similar views when he said that "*I cannot teach anybody anything. I can only make them think*".

Dale (1969) established the following relationships between the level of involvement of the learner and how much the learner remembers: "*We remember 10% of what we read, 20% of what we hear, 30% of what we see, 50% of what we hear and see, 70% of what we say 90% of what we do*".

Studies show that classroom attention-span of students varies from 15 to 25 minutes (Bligh, 1972 as cited by Cashin, 1985). Meyers & Jones (1993) found that students' classroom attention is poor for 40% of the time and they retain 70% of the information during the first ten minutes and only 20% of the information during the last ten minutes of a lecture. In a 1997 national survey in the USA, of more than 250,000 freshmen at nearly 500 universities, 35.6% of the students said that they were frequently bored in class (Berk, 2003).

All these facts stress the need of an active learning environment in particular when feedback from students has shown that a passive lecture in higher education is not a preferred method of learning (Sander et al., 2000 as cited by Huxman, 2005).

2.1 *Active learning: defined*

A literature review highlighted the following characteristics of active learning:

- Learning is enhanced by involving the student in activities and relationships, inside and outside the classroom (Astin, 1984);
- "Lectures are not well suited to higher levels of learning: application, analysis, synthesis, influencing attitudes or values. Lecturing is best suited to the lower levels of knowledge and understanding." (Cashin, 1985);
- "Learning is not a spectator sport. Students do not learn much just by sitting in classes listening to teachers, memorizing pre-packaged assignments, and spitting out answers. They must talk about what they are learning, write about it, relate it to past experiences and apply it to their daily lives." (Chickering & Gamson, 1987);
- Higher-order thinking through analysis, synthesis and evaluation is more important than reading, writing and discussing (Bonwell & Eison, 1991);
- "Active learning involves providing opportunities for students to meaningfully talk and listen, write, read, and reflect on the content, ideas, issues and concerns of an academic subject." (Meyers & Jones, 1993);
- "Although humans appreciate the common and the familiar, and often resist change, the brain seeks and reacts to innovative occurrences." (Angelo, 1991 as cited by Forrest, 2004);
- "In fact, rote learning (e.g., memorizing terminology) is frustrating because the brain resists meaningless stimuli." (Forrest, 2004).

Thus we may make the following definition:

Active learning involves designing, implementing, maintaining and promoting, within and outside classroom, environment for learning, through creating opportunities for active engagement with the subject matter. It strives for higher-order thinking and in-depth comprehension of the learner.

2.2 *Active learning: drawbacks*

Active involvement and interaction of the students translate into loss of teaching time as substantial classroom time is spent on multiple questioning and feedback sessions (Lammers & Murphy, 2002 as cited by Huxman, 2005), especially while handling large classes (Bonwell & Eison, 1991). It results in lesser content coverage in the classroom (Murray & Brightman, 1996 as cited by Huxman, 2005). But the biggest roadblock is a creation of the instructor himself and that is fear of failure, fear of loss of control due to enhanced classroom discussions, fear of lack of student participation, and fear of criticism of a new method (Bonwell & Eison, 1991).

2.2.1 *Handling drawbacks: short lectures*

An analysis of research papers on higher education by Chilcoat (1989) found that continuous rapid lecturing happens at a rate of 120 to 240 words per minute but most of the students can write down only 20 words per minute. The result is half-cooked notes and poor understanding during lecture time.

A solution to the above problem can be found in a study conducted by Ruhl et al. (1987), in which the instructor paused for two minutes for three times during a lecture, and during these pauses students worked in pairs to re-discuss the topic discussed by the instructor and then reworked on their class notes to fill up the gaps left while taking down the notes during the lecture. During the pauses students did not engage in discussion with the instructor. At the end of the lecture three minutes were given to let the students write down their free recall. Results proved that short lectures of 12 to 18 minutes followed by pauses for student interaction enhance students' learning.

Bonwell & Eison (1991) suggest to conduct guided lecture in which the instructor lectures continuously for 20 to 30 minutes without the students taking notes and then allots five minutes for free recall writing by the students followed by small groups of students discussing and analysing the topic.

2.2.2 *Handling drawbacks: revised lecture information density*

Russell et al. (1984) conducted research on 123 medical students and divided them into three groups with no significant difference in cumulative GPA's. The first group was exposed to a high density lecture with 90% new content; the second group was exposed to a medium density lecture with 70% new content; and the third to a low density lecture with 50% new content. In each lecture the remaining time was spent on reinforcing the core ideas by actively involving the students and by relating the content to prior experience. Tests after the lectures confirmed that learning and retention was higher with low density content.

Fedler & Brent (1996) suggest to discuss only core, critical and difficult topics in class, to give brief writing assignments to the students on the self covered topics and then to test the students on those topics. Wilke (2003) conducted research on 141 students in human physiology and found that active learning improved students' content achievement even when the content is reduced.

Research had proved time and again that reduced lecture time and reduced lecture information density enhances learning.

2.2.3 *Handling drawbacks: expectations and responsibilities*

Lesser lecture time in the classroom and lower density content should not lead to lower expectations of the students or the instructor. Chickering & Gamson (1987) suggest "*expect more and you will get more*". Seeler et al. (1994) suggest that an active learning environment does not mean structure less learning environment and it does not mean undermining the importance of theory. An ctive learning environment not only increases the opportunities for interaction and involvement but also responsibilities. Following are some important expectations and responsibilities as suggested by Seeler et al. (1994).

Students must be more aware and be ready to be responsible for their own professional development. They must increase their level of involvement in the active learning environment and focus on the application of theory to real life practice. But this does not mean that the instructor is less responsible for active learning environment. Seeler et al. (1994) further suggest active learning depends on the success of the instructor in helping the students successfully manage the transition from a passive style of learning to the new one. The instructor needs to understand the ability of the student, develop new skills, be aware of practical applications of theory to design exercises and assignments accordingly, adopt the role of facilitator and mentor. The instructor must rely on feedback and make changes, if necessary.

Stinson & Milter (1996) also highlight the instructor's skill development and student transition as two major pillars of active learning environment. They point out that the instructor needs to be an active listener and a coach, so that he can encourage the students to take responsibility for their own learning. This can be done effectively when the instructor gives continuous and prompt feedback to the students (Chickering & Gamson 1987).

2.2.4 *Handling drawbacks: student and instructor rapport*

Confucius believed in affection and empathy towards students (Huanyin, 1999). Chickering & Gamson (1987) advised instructors to "*encourage contact between student and faculty*". Tiberius &

Billson (Tiberius & Billson, 1991 as cited by Fleming, 2003) emphasized the need of a trust based, open and co-operative relationship between an instructor and his students which is guided, not by control, but by mutual agreement.

Active learning demands involvement. Involvement demands interaction. Fruitful interaction demands rapport. But that does not mean that an instructor has to indulge in students-appeasement. He has to rather strive for a growth-orientated relationship with sincerity and honesty.

2.3 *Co-operative active learning*

Active learning may be implemented through different routes. One of the popular routes is co-operative active learning. Students' involvement amongst themselves is crucial for successful active learning environment. As Chickering & Gamson (1987) stated, "*learning isn't a solo race rather it's a team effort which depends on collaboration and co-operation amongst students*". In co-operative learning a class is divided into smaller groups to enhance understanding and learning of students through positive interdependence, individual accountability, continuous interaction, teamwork, to accomplish orderly thinking, problem solving, critical analysis and clear expression (Johnson et al., 1998).

Imagine six stand-alone desktop computers and now compare them with a group of six networked computers. Which would give higher volume of fast information sharing and processing? Of course, the networked group. Co-operative learning aims at developing a network of human brains to achieve higher levels of learning, and to prevent this network from crashing the size of the teams is kept smaller. Student attendance is essential in a co-operative learning environment as absence of one or few team member(s) might disturb the task accomplishment and create a negative impact on the team spirit. (McManus, 1996).

Finlay & Faulkner (2005) conducted a project in which self help reading groups of 3–5 students were formed. Each student had to read a different topic each week and then had to write a one-page synopsis which was later discussed with the group. After the reading students asked questions among themselves; 73% of the students expressed positive perception of the method. Boud et al. (1999) refer to "peer learning" in which students learn in a group without intervention by an instructor.

Co-operative learning aims at self help in pursuit of knowledge. It may partially involve the instructor or it may be practised without instructor's intervention. One logical adaptation would be to initiate co-operative learning with a higher degree of instructor's involvement and intervention and to gradually reduce the involvement after judging the degree of self sufficiency of the group.

Co-operative learning may promote social loafing. Latane et al. (1979) in their research found that individuals tend to reduce effort when performing in a group. This is called social loafing. Maniar (2002) researched team work in business and sports and found that individual input reduces with the increase in size of a team. To prevent social loafing he suggests to evaluate each individual on his/her contribution, give smaller awards along the way instead of rewarding the group at the end of the task.

2.4 *Assignment-based active learning*

Since 1988, the University of Hawai at Manoa has been running "writing intensive" classes which aim at "learning through writing" (UHM 2006). An internal study has found that students have positive perception of the programme and writing intensive assignments enhance learning amongst the students, but they need to be taught how to write to learn. The strategy used by the university is to guide students to improve their analytical skills through synthesis of the journal articles, improve their reading skills through writing summaries, learn data collection techniques, to improve their analytical skills by examining research articles and then to present their own views through writing intensive assignments.

Plagiarism may be cited as a big drawback of writing intensive assignments but software like 'Turnitin' (www.turnitin.com) helps to keep a check on students indulging in plagiarism.

2.5 *Problem-based active learning*

Problem-based learning is a self-directed effort of a student towards understanding the solution of a problem. In this process the student identifies what he knows and what he needs to know to solve the problem (Barrows and Tamblyn 1980).

Instead of feeding the students with readymade content through lectures and discussions, problem-based learning begins with a challenge of presenting a real-life problem through case studies, research papers, assignments, etc., to the students, the answer to which they have to find on their own, individually and/or collectively. This method drastically changes the role of the students as well as the instructor.

The usual practice of teaching is to explain the concept to the students and then follow with questions. Problem-based learning (PBL) is exactly opposite as in PBL the beginning is with a problem and in due course of time students acquire knowledge and skills to solve the problem. A teacher is no more a provider of quick and easy answers; rather he becomes a facilitator in search of relevant information. The biggest challenge for successful implementation of PBL is the willingness and ability of the instructor to create a learning environment which is not teacher-centred and in which the student actively guides his own learning, individually or in a group (Barrows & Tamblyn, 1980).

The second most important challenge while implementing PBL is the selection of a problem. Stinson & Milter (1996) suggest the selected problem should reflect real life circumstances and be contemporary to facilitate the process of learning and doing. They further suggest the problem to be ill-structured, as real life problems are rarely structured. An ill-structured problem would enable the student to comprehend ill-defined situations. Duch (1996) insists that an effective problem must engage and motivate students, should be connected to previously learned knowledge, be open ended and not limited to one correct answer, must encourage a higher level of learning by asking for solutions based on facts and logic. Duch (1996) further clarifies that instead of dividing the problem into smaller units to be individually solved by a student, an effective problem should encourage total involvement of a students' group and prevent "divide and conquer" of the assignment. We may conclude that PBL attempts to bridge the gap between education and research.

3 MULTIPLE INTELLIGENCE AND MULTI-SENSE LEARNING

Gardner's (1993) concept of multiple intelligence identified that each individual has eight facets of intelligence. Logical-mathematical intelligence makes a learner quick with numbers, strong in reasoning and scientific thinking. Verbal-linguistic makes a learner comfortable with languages and communication including verbal and written assignments, etc. Musical-rhythmic intelligence relates to recognition and use of sounds and audio presentations. Naturalist intelligence relates to the ability to work with nature, field trips, real-life situations. Visual-spatial relates to the sense of sight and the ability to make mental images and concept maps. Inter-relational relates to interpersonal skills such as communication and team work. Intra-relational is associated with understanding of self. Bodily-kinaesthetic is associated with body movements.

Visua- based instructions using audio-visual models, films, 3D graphs enhance the productivity of an active learning environment (Cashin, 1985; Bonwell & Eison, 1991). VARK (Visual, Aural, Read/Write, Kinaesthetic) learning strategy attempts to create an active learning environment. In visual learning an instructor uses gestures and picturesque language, pictures, videos, posters, slides, colour flow-charts, colour graphs and diagrams, underlining and different colour highlighters, etc.

In aural learning, classroom discussions between learner and instructor and amongst learners play an important role. Outside classroom strategy involves spending time in a quiet place and recalling the ideas and speaking out the answers aloud or inside the head to register the content in mind.

Read/write learning strategy works well when the content involves lots of definitions, glossaries, manuals, dictionaries. This strategy requires the students to undertake multiple writing and reading sessions. To reinforce learning frequent multiple-choice tests are suggested.

In the kinaesthetic learning strategy, field tours and trips, lectures with real-life examples, laboratory sessions, learning from trial and error etc are given emphasis. The use of case studies and applications to help with principles and abstract concepts is suggested. (Complete information on VARK adapted from www.vark-learn.com.)

A comprehensive active learning environment which can successfully engage the learners with consistency will need to address the issues of multiple intelligence and multiple human senses. A multi-dimensional communication will strengthen the learning environment. Multiple intelligence identifies different facets of human intelligence, and a successful active learning environment needs to implement a variety of tools to address these facets.

4 CONCLUSION

Active learning is more than a technique or a tool. Active learning is a way of life. There are three major components of an active learning environment and they are students, instructors and institute/management. Active learning environment demands behaviour modification on the part of students, instructors and institute/management. The speed and intensity of implementing a step-by-step comprehensive active learning environment in higher education has to be determined by the desire, ability and willingness of the students, instructors and institute/management.

Instead of developing and implementing a learning environment which is decided by the management/faculty/consultant combination, the need for identification for an activity-based environment in the mindset of the students has to be given importance. There is a need of a paradigm shift from the usual approach of 'how I like to teach' to 'how the learner wants to learn and how the learner will learn better'.

Apart from the methods already discussed there are multiple tools to choose from for implementing an active learning environment, such as,

- Group brainstorming;
- Think-pair-share (a tool in which students are divided into groups of two and they discuss their understanding of the topic after a short lecture);
- One-minute papers on the muddiest point, difficult topic, easiest definition etc.;
- Concept maps (a tool which abets in comprehending the logical relationships amongst a number of ideas within a topic of discussion by creating a visual map of the connections);
- Students' role play, simulations;
- Short writing assignments/summaries;
- Classroom debates and discussions.

One can pick and choose any combination of the methods and tools discussed in this paper but the selection of a combination has to be guided by the habits, preferences, willingness, behaviour modification, knowledge base and learning style of the students and also the instructors.

The following may be considered as prerequisites for a successful active learning environment:

1. Clarity, creativity and likeability of the tools and techniques used;
2. Degree (positive) of behaviour modification (engagement and involvement) of the students induced by the tools used;
3. Synergy of active learning environment with the present and required knowledge base of the instructor and learners;
4. Synergy of active learning with course, syllabus and time schedule;
5. Availability of relevant case studies, problems, assignments, tests, related infrastructure, etc.;
6. Complete involvement of all the stakeholders: faculty, support staff, institute's management, students (individually and jointly) and parents;
7. Examination performance enhancement by the tools and techniques used;
8. Within and outside the classroom as well as within and outside college, learning continuity.

To conclude, active learning would start with an instructor who is an active learner and active instructor himself and who can upgrade his knowledge and skills in the time of need. He has then to identify the needs of the students with reference to changes expected in the learning environment. He then has to balance the students' needs with what he thinks is best for the learning environment. This would lead to the preliminary preparations in collaboration with the learners and would be followed by gradual, step-by-step implementation of the new methods and tools.

Continuous feedback from the students and supervision by the instructor might be essential at an early stage. Regular evaluation of the learning environment is essential through class tests, *viva voice*, evaluation of assignments (by the instructor or by the students themselves), students' feedback (informal classroom feedback and/or formal questionnaire-based feedback), peer feedback (by instructor's colleagues).

ACKNOWLEDGEMENTS

The author expresses his gratitude to Dr Praveen Dhyani, Director, BITIC, Ras Al Khaimah, UAE, for his continuous guidance.

REFERENCES

Astin, A.W. 1984. Student involvement: A developmental theory for higher education. *Journal of College Student Personnel*, 25: 297–308.

Barrows, H.S. & Tamblyn, R.M. 1980. *Problem-based Learning: An Approach to Medical Education*. New York: Springer

Berk, R.A. 2003. Professors are from Mars, Students are from Snickers: How to write and deliver humor in the classroom and in professional presentations. Sterling, V.A.: Stylus Publishing.

Bligh, D.A. 1972. What's the use of lectures? Harmondsworth, England: Penguin Books.

Bonwell, C. & Eison, J. 1991. Active learning: Creating excitement in the classroom. ASHE-ERIC Higher Education *Report* No. 1, George Washington University School of Education and Human Development, Washington, D.C.

Boud, D., Cohen, R. & Sampson, J. 1999. Peer learning and assessment. *Assessment in Higher Education*, 24: 413–26.

Cashin, W.E. 1985. Improving Lectures, *Idea Paper* No 14. Center for Faculty Evaluation and Development, Kansas State University, Manhattan, KS.

Chickering, A.W. & Gamson, Z.F. 1987. Seven principles for good practice. *AAHE Bulletin*, 39(7): 3–7.

Chilcoat, G.W. 1989. Instructional behaviors for clearer presentations in the classroom. *Instructional Science*, 18: 289–314. Abstract of the research findings downloaded on 10 June 2006 from http://www.active-learning-site.com/sum1.htm

Dale, E. 1969, *Audio-Visual Methods in Teaching* (3rd Edition). London: Holt, Rinehart, and Winston.

Duch, B. 1996. Problems: A Key Factor in PBL. Center for Teaching Effectiveness. Downloaded on 15 June 2006 from http://www.udel.edu/pbl/cte/spr96-phys.html

Fedler, R.M. & Brent, R. 1996. Navigating the bumpy road to student-centered instruction. *College Teaching*, 44: 43–47.

Finlay, S.J. & Faulkner, G. 2005. *Tete-à-tete*: Reading groups and peer learning. *Active Learning in Higher Education*, 6(1): 32–35.

Fleming, N. 2003. Establishing Rapport: Personal Interaction and learning, *Idea Paper* No. 39. Center for Faculty Evaluation and Development, Kansas State University, Manhattan, KS.

Forrest, S. 2004. Learning and teaching: The reciprocal link. *The Journal of Continuing Education in Nursing*, 35(2): 74–79.

Gardner, H. 1993. *Multiple Intelligences: The Theory in Practice*. New York: Basic.

Huxman, M. 2005. Learning in lectures: Do 'interactive windows' help? *Active Learning in Higher Education*, 6(1): 17–31.

Huanyin, Y. 1993. Confucius: K'ung Tzu (551-479 BC). *Prospects: The Quarterly Review of Comparative Education* (Paris, UNESCO: International Bureau of Education), vol. XXIII, no. ½, pp. 211–19.

Johnson, D.W., Johnson R.T. & Smith, K. 1991. *Active Learning: Co-operation in the College Classroom*. Edina, MN: Interaction Book Company.

Lammers, W.J. & Murphy, J.J. 2002. A profile of teaching techniques used in the university classroom. *Active Learning in Higher Education*, 3: 54–67.

Latane, B., Williams, K. & Harkins, S. 1979. Many hands make light the work: The causes and consequences of social loafing. *JPSP*, 37(6): 822–832.

Maniar, S. 2004. Improving group performance in business & sport: Part I—Social loafing. *Optimal Performance Newsletter*, 2(2).

McManus, D.A. 1996. Changing a course from lecture format to co-operative learning. *Paideia: Undergraduate Education at the University of Washington*, 4(1): 12–16.

Meyers, C. & Jones, T.B. 1993. Promoting Active Learning: Strategies for the College Classroom. San Francisco: Jossey-Bass.

Murray, R. & Brightman, J.R. 1996. Interactive teaching. *European Journal of Engineering Education*, 21: 295–308.

Ruhl, K.L., Hughes, C.A. & Schloss, P.J. 1987. Using the pause procedure to enhance lecture recall. *Teacher Education and Special Education*, 10: 14–18. Abstract of the research paper downloaded on 10 June 2006 from http://www.active-learning-site.com/sum1.htm

Russell. I.J., Hendricson, W.D. & Herbert, R.J. 1984. Effects of lecture information density on medical student achievement. *Journal of Medical Education*, 59: 881–889. Abstract of the paper downloaded on 10 June 2006 from http://www.active-learning-site.com/sum1.htm

Sander, P., Stevenson, K., King, M. & Coates, D. 2000. University students: Expectations of teaching. *Studies in Higher Education*, 25: 309–323.

Seeler, D.C., Turnwald G.H. & Bull, K.S. 1994. From teaching to learning: Part III. Lectures and approaches to active learning, *Journal of Veterinary Medical Education*, 21(1). (Spring). Downloaded on 15 June 2006 from http://scholar.lib.vt.edu/ejournals/JVME/V21-1/Seeler1.html

Stinson, J.E. & Milter, R.G. 1996. Problem-based Learning in Business Education: Curriculum Design and Implementation Issues. New Directions in Teaching and Learning in Higher Education. San Francisco: Jossey-Bass. Downloaded on 15 June 2006 from http://www.ouwb.ohiou.edu/stinson/PBL.html

Tiberius, R.G. & Billson, J.M. 1991. The social context of teaching and learning. in: *College Teaching: From Theory to Practice*, Menges, R.J. and Svinicki, M. (eds), San Fancisco: Jossey-Bass.

Wilke, R.R. 2003. The effect of active learning on student characteristics in a human physiology course for non-majors. *Advances in Physiology Education*, 27: 207–223.

UHM, 2006. *Writing Matters, Manoa Writing Program*. The University of Hawai at Manoa. Downloaded on 25 June 2006 from http://www.mwp.hawaii.edu/resources/wm1.htm

Higher Education in the Twenty-First Century: Issues and Challenges – Al-Hawaj, Elali & Twizell (eds)
© 2008 Taylor & Francis Group, London, ISBN 978-0-415-48000-0

The impact of blended learning to enhance the quality of higher education

Saravanan Nallaperumal & Shanthi Saravanan
Birla Institute of Technology, Kingdom of Bahrain

ABSTRACT: Educational systems around the world are under increasing pressure to use the new Information and Communication Technologies (ICT) to teach students the knowledge and skills they need in the 21st century. Organizations today are looking beyond the automation of traditional training models to new approaches to knowledge transfer and performance support that are better aligned with business goals and deliver measurable results. By focussing on the specific business objective, rather than the learning technology, we are given the opportunity to fundamentally re-think how we design and deliver learning programmes. In this paper we consider the potential impact of blended learning in higher education with regard to pedagogy in the teaching and learning process and how the quality of higher education could be achieved by integrating it into the teaching and learning process in higher learning institutions. It also discusses the key issues of quality of higher education and how quality can be improved through blended learning modes with an analysis of some case studies.

1 INTRODUCTION

What will education, especially higher education, be like in the 21st century? How will blended mode impact the delivery of instruction in the new millennium? We keep asking ourselves these questions. In today's knowledge-driven era, higher education is a precondition to full and satisfying participation in the global economy and society. Globally, the rate of participation in higher education continues to rise. Today's colleges and universities continue to prosper and adapt by letting many flowers bloom. Differing institutional structures have emerged to address the changing educational and economic needs of changing populations. New information technologies are developed and adapted to serve differing learning styles. Higher education institutions have moved far away from the ivory tower and now extend into the very fabric of the diverse societies in which they operate (Educase, 2005). Educators who have tried both approaches realize that neither the traditional lecture format nor the distance education approach is appropriate for every student, every teacher, and every course. This leads us to a third option, blended learning (Franks, undated). The multimedia and digital technologies that will play a role in the classrooms of the future are many and varied (Wang, undated).

Traditional physical classrooms have been the dominant form of knowledge transfer for at least 3,000 years. Even today, nearly 80% of corporate training is conducted in the classroom (ASTD, 2001). The last universal technology in learning, the printed book, is over 500 years old. Yet in the past 10 years alone, over 10 major new technologies for learning and collaboration have been introduced. Early experience with these technologies has uncovered opportunities for profound improvements in quality, effectiveness, convenience and cost of learning experiences. Only now are we beginning to understand how learning experiences will evolve to exploit "blended" combinations of both traditional and technology-based learning methods, and how blended learning can have a strategic impact on critical business processes (Singh & Reed, undated).

2 THE CHALLENGE OF QUALITY IN HIGHER EDUCATION

The journey of self-reflection that leads to a quality-driven organization is a struggle for higher education because of its long history and traditions (Thorne, 2003). In an attempt to ensure that higher education's customers are satisfied, and to take the pressure off the academy from its external stakeholders, many institutions have begun to implement quality initiatives. If leaders within higher education follow the points of the quality gurus, like W. Edwards Deming, Philip Crosby, Joseph Juran, and Imani, they will have:

- constructive competition,
- shared values and unity of purpose,
- collaboration on broad issues,
- simulations and synergistic planning,
- emphasis on responsibility to contribute,
- decentralized partnerships built upon situational management,
- team accountability,
- constancy of purpose,
- win-win resolution to conflicts via conflict management,
- probably most important, a superior professorate, student body, and administration.

In summary, the organizational culture will be transformed and front-line employees will become a part of implementation of organizational goals.

3 QUALITY IMPROVEMENT PROCESS

The following management recommendations need attention when implementing the continuous quality improvement process:

- develop top leadership commitment,
- recognize quality improvement as a system,
- define it so that others will recognize it,
- analyse change behaviour that leads to service,
- measure the quality of the system,
- develop improvements in the quality of the system,
- define who the customers are,
- measure the gains of quality and link them to customer satisfaction,
- develop plans to ensure constant results,
- attempt to replicate the improvements in other areas of the organization.

The university's passionate convictions and traditional ways of doing business and its tightly held assumptions about quality make it difficult to incorporate a shift in the way it does business.

4 *ICT* IN HIGHER EDUCATION

ICT has the potential to transform the nature of education: Where and how learning takes place, and the roles of students and teachers in the learning process. The main challenge confronting our educational systems is how to transform the curriculum and teaching-learning process to provide students with the skills to function effectively in this dynamic, information-rich, and continuously-changing environment. ICTs provide an array of powerful tools that may help in transforming the present isolated, teacher-centred and text-bound classrooms into rich, student-focussed, interactive knowledge environments. To meet these challenges, learning institutions must embrace the new technologies and appropriate ICT tools for learning. The learning and teaching methodology may mean that one or the other information and communications technology has to be selected.

Table 1. Technologies and communication quality issues.

Technology	Person to machine	One to one	One to many	Many to many	Same time	Different time	Symmetric	Asymmetric
Audio tapes	●	○	●	●	○	●	○	●
Radio	○	○	●	●	●	○	○	●
Telephone	○	●	○	○	●	○	●	○
Audio-conferencing	○	○	●	●	●	○	●	○
Video tapes	●	○	●	●	○	●	○	●
Television broadcasts	○	○	●	●	●	○	○	●
Videoconferencing	○	●	●	●	●	○	●	○
Resource search on WWW	●	●	○	○	○	●	○	●
Publishing on the WWW	●	○	●	●	○	●	○	●
Using CD-ROMs, DVDs	●	●	●	●	○	●	○	●
Email	○	●	●	●	○	●	●	○
Computer conferencing	○	●	●	●	○	●	●	○
Mailing lists	○	○	●	●	○	●	●	○
Electronic Learning Platforms	●	●	●	●	○	●	○	●
Online simulations	●	●	○	○	●	○	●	○
Online laboratories	●	●	○	○	●	○	●	○
Computer Apps (eg Word, Powerpoint)	●	●	○	○	●	○	●	○
Computer games	●	●	●	●	●	○	●	○
Educational software	●	●	○	○	●	○	●	○

○ Not supported ● Not inherently supported ● Fully supported

(*Source*: JISC Report: Satellite Applications in Education: Education content: Contribute, Distribute and Retrieve.)

Table 1 gives an indication of how each of the most common used technologies specifically addresses the communication quality issues of topology, synchronicity and symmetry.

According to Michael Albright, there are seven functional areas of instructional technology in higher education and they are: instructional development, learning resources, classroom technologies, media development, instructional telecommunication, academic computing, and research and evaluation (Wang, undated).

Hence Blended Learning = Face-to-face Learning + ICT

5 PERSPECTIVES OF BLENDED LEARNING

Blended learning focusses on optimizing achievement of learning objectives by applying the "right" learning technologies to match the "right" personal learning style to transfer the "right" skills to the "right" person at the "right" time (Singh & Reed, undated.)

In higher education, blended learning is often referred to as a hybrid model. The goal of this hybridization is to join the best features of in-class teaching with the best features of on-line learning to promote active, self-directed learning opportunities for students with added flexibility (Garnham & Kaleta, 2002). The following are all types of blended learning:

- A mix of media and tools in an e-learning environment, for example video, audio, text, computer-assisted learning (CAL) packages, etc.,
- A hybrid approach that combines face-to-face instruction and online activity,
- A combination of self-paced and group-based work, or a combination of instructional approach, using on-line tools.

To create a successful blended-learning approach to teaching and learning, two tools are essential: 1) learning technology and 2) blended learning strategy (McArthur, 2001).

Learning technology enables the instructor to create course materials in different formats including printed textbooks, on-line textbooks, on-line lecture notes, QuickTime movies, CD ROMs, PowerPoint™ presentations, streaming media, and videotapes of lectures and demonstrations. Dr David Rose, co-Director of the Centre for Applied Special Technology (CAST), believes in providing students with "multiple representations of information" and a menu of different ways to get the information they need (O'Neill, 2001).

The original use of the phrase "Blended Learning" was often associated with simply linking traditional classroom training to e-Learning activities. However, the term has evolved to encompass a much richer set of learning strategy "dimensions." Today a blended learning programme may combine one or more of the following dimensions, although many of these have over-lapping attributes (Singh & Reed, undated).

6 INGREDIENTS OF THE BLEND

Today organizations have myriad learning approaches to choose from, including but not limited to the following:

6.1 *Synchronous physical formats*

- Instructor-led classrooms and lectures,
- Hands-on laboratories and workshops,
- Field trips.

6.2 *Synchronous on-line formats*

- E-meetings,
- Virtual classrooms,
- Web seminars and broadcasts,
- Coaching,
- Instant messaging.

6.3 *Self-paced, asynchronous formats*

- Documents and web pages,
- Web/Computer-based training modules,
- Assessments/Tests and surveys,
- Simulations,
- Job aids and electronic performance support systems (EPSS),
- Recorded live events,
- On-line learning communities and discussion forums.

According to Bonk & Graham (in press), the expected growth of Blended Learning in Higher Education Settings is as shown in Figure 1.

7 BLENDED LEARNING CHALLENGES

There is a saying in the e-learning world that "if content is king, then technology is God." This holds true when it comes to blended learning as well. The technology skill level of learners and facilitators can also be a key challenge for blended learning. Providing adequate technical support and training to participants and facilitators is critical to successful blended learning. Blended

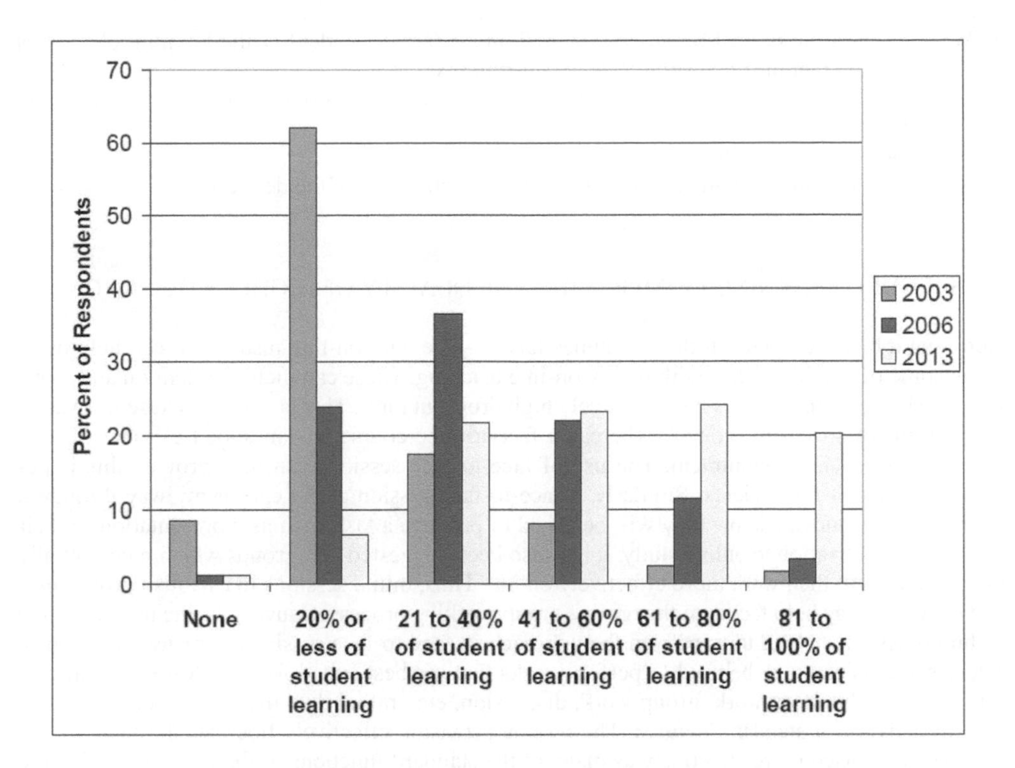

Figure 1. Expected Growth of Blended Learning in Higher Education Settings.
(*Source*: Bonk & Graham, in press).

learning places more responsibility for learning in the hands of the learner and, for some learners not accustomed to independent learning, this can be challenging. Building in support for learners from instructors, supervisors and peers helps to increase course completion rates and learner satisfaction. Blended learning also requires an intentional approach to instructional design so that the programme is blended in design, not just in delivery. Professional development for instructors to learn online teaching strategies and facilitation skills is important to any successful blended programme. Instructors must also be familiar with all of the technologies that will be used in a blended programme and be able to support learners. Finally, instructors must also learn to integrate methods of assessing student learning within the context of blended learning. Creating a successful blended learning programme requires thoughtful analysis and design, just like any training programme. To be successful, both the design and the implementation of blended learning needs to be intentional (Staley et al., 2007).

8 CRITERIA FOR A SUCCESSFUL BLENDED-LEARNING PROGRAMME

In reality the underpinning principles of blended learning are no different from any other form of learning (Thorne, 2003). The key criteria are based on the following:

1. Identifying the core learning need,
2. Establishing the level of demand/timescale,
3. Recognizing the different learning styles,
4. Looking creatively at the potential of using different forms of learning, i.e., matching the learning need to different delivery methods and identifying the best fit,

5. Working with the current providers, internal and external, to identify the learning objectives and to ensure that the provision meets the current need,
6. Undertaking an education process and developing a user-friendly demonstration to illustrate the potential of blended learning,
7. Being prepared to offer follow-up coaching support,
8. Setting up a monitoring process to evaluate the effectiveness of the delivery.

9 BLENDED LEARNING IN IMPROVING THE QUALITY OF HIGHER EDUCATION

Adopting a blended approach that combines face-to-face with on-line instruction can help overcome some of the difficulties with purely on-line teaching. These can include retention and motivation. Many on-line courses have relatively high drop-out rates. This is partly because it is easier to drop out of an on-line course, whereas a face-to-face course has an added element of peer pressure and social commitment. The use of face-to-face sessions can help provide this impetus and cohesion for students. Similarly, a face-to-face session at the end or midway through a course, where students know they will be asked to perform a task such as a presentation to their peers, adds motivation to online study. It has also been suggested that groups which meet initially face-to-face will then work more effectively online. Thus, initial sessions to establish group roles and cohesion may help facilitate the on-line group working process. Conversely, the use of on-line technologies has helped to transform the approach in face-to-face sessions in many universities. These sessions can now be used to perform tasks that are best suited for face-to-face situations, for example laboratory work, group work, discussion, etc., rather than the straightforward information delivery of standard lectures. The on-line provision effectively liberates the face-to-face sessions to be used more effectively as many of the standard functions of the face-to-face lecture can now be accomplished on-line (Weller, undated).

10 SUPPORTING CASE STUDIES

The following case studies are evident to illustrate the role of blended learning to improving the quality of higher education:

Case 1: Portals are taking off on campuses everywhere. According to Campus Computing 2006, the Campus Computing Project's survey of 540 two- and four-year public and private colleges and universities across the US, portal deployment for four-year public residential universities jumped from 28 per cent to 74 per cent of responding institutions between the 2002 and 2006 academic years; from 20 per cent to 38 per cent for private four-year universities; and from 23 per cent to 43 per cent for community colleges. Yet, as more and more campuses buy into the promise of single sign-on and integration of information and services, needs and realities of implementation can diverge. To get a better sense of the "state of the portal" in higher education, it is important to look at the differing stages of maturity and the wide range of technology choices in the portal journey.

http://campustechnology.com/articles/39380_2/

Moreover, the surviving and thriving online programmes help expand access to higher education, a key goal of public policy for the last 40 years. Unknowingly acknowledging the preamble to the historic Higher Education Act of 1965, the new dotcoms—public, private, and for-profit—serve "all who might benefit" from the opportunity to attend college.

Case 2: According to the 2006 Campus Computing Survey, wireless networks now reach half of college classrooms, compared to just over two-fifths in 2005, and a third in 2004. In addition, more than two-thirds of campuses participating in the annual survey have a strategic plan for deploying wireless as of the fall of 2006, up from 64 per cent in 2005, and more than 53 per cent in

2004. Results from the 2006 survey reveal that three-fifths of colleges and universities increased their campus IT budgets for wireless for the academic year (Slabodkin, undated).

Case 3: There is a suggestion that a blended approach achieves better learning outcomes. In 2002, Thomson Learning (who own the e-learning providers netG) reported the findings of a survey where 128 employees from a range of organizations were part of an experiment that compared blended learning with single delivery on-line courses, focussing on learning the spreadsheet package Excel. In this case, a blended learning course was one which was 'based on scenario-based exercises aligned with learning objectives and integrated with the use of actual software, live on-line mentoring, and other support material'. The blended learning approach demonstrated a 30 per cent increase in accuracy of performance and a 41 per cent increase in the speed of performance in the use of the software over the on-line course (Weller, undated).

Also, a study on the benefits and challenges of blended learning in higher education from the perspective of students, faculty and administration done by Norman Vaughan (Vaughan, 2007) reveals that students indicated that a blended learning model provides them with greater time flexibility and improved learning outcomes. Faculty suggested that blended courses create enhanced opportunities for teacher-student interaction, increased student engagement in learning, added flexibility in the teaching and learning environment and opportunities for continuous improvement; and from the administrator's point of view, blended learning presents the opportunity to enhance an institution's reputation, expand access to an institution's educational offerings, and reduce operating costs.

11 CONCLUSION

A recent survey of e-learning activity found that 80 per cent of all higher education institutions and 93 per cent of doctoral institutions offer hybrid or blended learning courses (Arabasz et al., 2003). The process of education and training is a fascinating and constantly changing journey. It requires an understanding of people, processes, technology and culture. Blended learning environments (BLEs) promise to be an important part of the future of both higher education and corporate training. Over the past decade, with the increased availability of technology and network access, the use of BLEs has steadily grown. Technology, training and support should be available for students and professional development support for the faculty. However, the amount of research done related to the design and use of BLEs is relatively small and additional research is needed.

REFERENCES

Arabasaz, P., Boggs, R. & Baker, M.B. 2003. Highlights of e-learning support practices. *Educase Centre for Applied Research Bulletin*, 9: 1–11.

ASTD, 2001. State of the Industry Report, *American Society for Training & Development*, March 2001.

Bonk, C.J. & Graham, C.R. (eds) (in press). Handbook of Blended Learning: Global Perspectives, Local Designs. San Francisco, CA: Pfeiffer Publishing.

Educase, 2005. *The Pocket Guide to US Higher Education*, 2005.

Franks, P.C. Blended Learning: What is it? How does it impact student retention and performance? World Conference on e-learning:1480–1482.

Garnham, C. & Kaleta, R. 2002. Introduction to hybrid courses. *Teaching with Technology Today*, 8(6). Source available from http://www.uwsa.edu/tt/articles/garnham.htm

McArthur, J. 2001. Blended learning: a multiple training strategy. Video Webcast. [On-line]. Available: http://www.connectlive.com/events/opm/.

O'Neill, L. 2001. Universal design for learning: making education accessible to all learners. *Syllabus*, 14(9): 31–32.

Singh, H. & Reed, C. *A White Paper: Achieving Success with Blended Learning*, Centra Software.

Slabodkin, G. Wireless Takes American Campuses by Storm.

Staley L. et al. 2007. *Blended Learning Guide*, Web Junction.
Thorne, K. 2003. Blended Learning: How to Integrate Online and Traditional Learning. Kogan Page.
Vaughan, N. 2007. Perspectives on blended learning in higher education. *International Journal on E-learning*, 6(1): 81–94.
Wang, H. Multimedia and Digital Technology: Challenge or Opportunity For Instructional Media Services in Higher Education?
Weller, M. Blended learning. Milton Keynes, UK: The Open University http://intranet.open.ac.uk/wdc-aps/sites/m2005_5_26_52124/s2005_11_16_39130/objects/d1677.mht

Staley L. et al. 2007. *Blended Learning Guide*, Web Junction.
Thorne, K. 2003. Blended Learning: How to Integrate Online and Traditional Learning. Kogan Page.
Vaughan, N. 2007. Perspectives on blended learning in higher education. *International Journal on E-learning*, 6(1): 81–94.

Higher Education in the Twenty-First Century: Issues and Challenges – Al-Hawaj, Elali & Twizell (eds)
© 2008 Taylor & Francis Group, London, ISBN 978-0-415-48000-0

Revitalizing research in Gulf universities through the scholarship of teaching and learning

Sofiane Sahraoui
Brunel University, Uxbridge, Middlesex, UK and American University of Sharjah, UAE

ABSTRACT: The competitive knowledge market is pushing for a differentiation in the higher education (HE) field whereby research is concentrated in elite universities and teaching constitutes the core mission of a larger number of HE providers (Lucas, 2006). Whilst such a strategy of differentiation could be viable, though debatable, in large HE sectors such as the UK's, it carries serious risks for institutions in the Gulf where the research capacity of HE institutions is still limited. The scholarship of teaching and learning, first advocated by Boyer (1990), and subsequently developed as a main agenda by the Carnegie Academy for the Scholarship of Teaching and Learning (CASTL) appears hence as a promising third way for the local HE sector breaking away from the tired teaching *versus* research debate. It seeks to integrate teaching and research by advocating several scholarships of teaching, integration (e.g., writing textbooks), and service alongside the scholarship of discovery or what is commonly recognized as conventional research. This paper will highlight the debate of teaching *versus* research with its implications for the Gulf HE scene, and introduce the scholarship of teaching and learning as a viable and promising alternative to build solid research capacity in the region. With Gulf universities called upon to educate a huge influx of students, teaching appears as the core mission of local universities hence warranting research on teaching and learning from within the disciplines. This would enable HE institutions in the Gulf to spread their scholarship resources based on their core mission rather than in the pursuit of an ever elusive excellence in research.

1 INTRODUCTION

Universities have acquired a highly significant role because of their perceived role as the front line within the knowledge-creating society (Slaughter & Leslie, 1997). In countries where the knowledge market is highly developed, they are engaged in a system of 'academic capitalism'[1] aiming to better serve the needs of industry in terms of science and technology innovation (Lucas, 2006). Whilst such integration to the knowledge market ensures pre-eminence, prestige, and resources for universities, this situation is not the lot of many institutions of higher education (HE). Rather, most perform remotely from the knowledge market and come closest through graduates rather than research output. It becomes thus necessary that the primary output of universities is of a good quality. Yet, and despite an increasing trend to ascertain the quality of academic delivery through accreditation and external evaluation, little is done in the way of researching and critically examining teaching and learning, the two determinant processes in ensuring a requisite quality among student graduates. This paper will first outline the increasing global differentiation in the HE sector with a special emphasis on the UK system, and introduce the scholarship of teaching and learning as a valuable and more accessible form of scholarship that would enable universities in the Arab region to claim a stake amongst institutions with respectable academic capital.

[1] Concepts of academic capitalism, research capitalism, and other forms of symbolic capitalism are commonly used in the social science literature as adaptations of Bourdieu's (1988) thinking tools.

In the UK, the HE Act of 2004 seems to be ushering a reinforced stratification in English HE, with a reversal to a split between vocational and academic training on the one hand and a distinction between research and teaching on the other (Deem, 2006). By absolving institutions from holding research degree granting powers before applying for university status, the state laid the ground for the development of a HE market primarily servicing the teaching needs of a massive influx of students. This materialized in an earlier assertion by HEFCE[2] that it is not appropriate for all universities to conduct research (Deem, 2006). Whilst most institutions still cling on the hope of playing the research game, as Lucas (2006) calls it, and accumulate a critical mass of research capital to place them in a competitive position in the global market, the increasing selectivity of research funding, both public and private, on the one hand and the rising costs of research on the other, will convince many to withdraw from the research game after a few unsuccessful RAE[3] bids. The RAE score will increasingly become a stigma for those failing to perform well. The changes introduced in the grading scheme of the 2008 RAE are meant amongst other things to disentangle the concentration of RAE star institutions in the upper echelons and create variation in the sector so that research funding can be further concentrated within fewer institutions. Already, nearly half the number of institutions that entered the last RAE secured less than 5% of their funding from the ensuing QR[4] funding whereas less than 1% availed of more than half of their funding from the same exercise (Lucas, 2006, p. 30) and this polarization is expected to further increase especially at the expense of post-92[5] universities which are under-represented in the RAE panels (Sharp, 2005). The concentration of research capital hence transcends individual academics[6] to cover HE institutions themselves in an attempt to structure HE along more efficient and transparent modes of delivery; and create two distinct types of institution, the research universities and the teaching universities; both with an important role to play in the knowledge economy and with research capital bestowing higher prestige on the research universities.

Humboldt's[7] dream of the university of scholarship (or *Wisenschaft*) where both *Lehrfreiheit* (freedom to teach) and *Lehrnfreiheit* (freedom to study) would be guaranteed does not the suit the knowledge economy and its neo-liberal precepts. Indeed, academic freedom goes against the spirit of market agency as knowledge is only legitimate if it has a buyer and not when it is pursued for its own sake. The synergetic relationship between research and teaching that Humboldt advocated (King, 2004) is no longer suitable from a market point of view and a new relationship of antagonism is emerging between teaching and research with the long term risk of a full rupture between pedagogy and research (Barnett, 2006). This fragmentation of the role of academics into teachers and researchers is as damaging to research as it is to teaching (Naidoo, 2004). Market efficiency and the neo-classical principle of the optimal allocation of resources call for the specialization of 'economic operators' in what they do best, research for some and teaching for others. Nor does Newman[8] fare any better though he could accommodate well with a separation of teaching and research. Teaching itself is being subjected to criteria of usefulness and utilitarianism going

---

[2]Higher Education Funding Council of England.

[3]Research Assessment Exercise is conducted every few years to rate UK institutions in terms of research performance. It is also determinant, as it forges perceptions of excellence in research, or lack thereof, among other donors such as the research councils, private charities, and corporate clients.

[4]Quality-Related.

[5]Post-92 universities are the former polytechnics which were transformed into universities to create a competitive market of HE.

[6]The research faculty *versus* the teaching faculty is a recurrent theme especially among universities with a dual mission of teaching and research.

[7]Ludwig Van Humboldt is founder of the University of Berlin and the father of the German university whose model was adapted during the establishment of the first land-grant institutions in the USA.

[8]Reverend John Newman was schooled and taught at Oxford in the 19th century before establishing the Catholic University of Dublin. He is considered by many as the forefather of modern liberal education.

against the philosophy of Newman for universal knowledge; one that is devoid of usability concerns and rather seeped into the 'cultivation of the mind' and the liberation of the 'gentleman' from earthly and material concerns.

The process of differentiation between universities is real and the prospects of a publicly-funded university specializing in both research and teaching becoming an endangered species is also real (Deem, 2006). Lucas recounts how a post-92 university is struggling to join the rank of research universities, improving its RAE score in 1996 though losing funding as a result of HEFCE's new allocation policy. Universities, even those with modest research bases, are trying to cling onto the endeavour of building a research capital, a pre-condition for international prominence, but chances are that most will fail and accept their new role as teaching institutions giving shape to a clearly stratified system of HE; one differentiated on the basis of teaching and research. Industry for its part is increasingly reinforcing policies of selectivity and concentration by investing large amounts of money in HE institutions with high research stature and reputation (Henkel, 2000). This new hierarchy of institutions with an increasing commoditization of knowledge in the teaching institutions could plunge the system of HE into severe inequalities with the lower-end HE institutions serving the under-privileged with knowledge that is devoid of significant "skills of innovation and ability to learn how to learn" (Naidoo, 2004). Faced with this inevitable stratification, HEFCE has pledged additional funding for the scholarship of teaching and learning, in an initiative called "a research-informed teaching environment" in those institutions that are in receipt of little pure research funding from the RAE (Deem, 2006). This would enable teaching institutions to develop a research base targeting their internal delivery process of teaching and learning rather than market-driven research.

In other parts of the world, including the Arab region, academic and research capital has been very slow to build and as the barriers to entry are further raised, it has become very difficult to develop world class institutions, at least from the standpoint of research. This does not mean, however, that such an objective is beyond reach as the example of the Hong Kong University of Science and Technology shows[9]. A more feasible strategy, however, is one based on accumulating academic capital through the scholarship of teaching and learning, which is more relevant to the reality of the bulk of HE institutions in the world. This is further elaborated below.

3 THE RELEVANCE OF SCHOLARSHIP

Whilst the global process of differentiation and diversification advances unabated within a firm global agenda, the situation is not doomed for universities that cannot compete on an equal footing with the large international universities that have mustered a large amount of research capital. On the contrary, the differentiation process will create educational niches for the bulk of HE institutions left out of the global research game. This niche is two pronged; one entails generating local knowledge as well as adapting global knowledge to local contexts and the second entails the development of scholarship of teaching and learning becoming all the more important for institutions whose core mission is teaching rather than research. To borrow a concept from the developmental economist Samir Amin[10], the research agenda in the so-called teaching institutions has to be 'delinked' from the global research agenda, for in the absence of a critical mass of research capital, research output in teaching institutions is bound to be of sub-standard quality and with virtually no impact on global research, except for a very few isolated contributions. The first form of scholarship, development of local knowledge and adaptation of global knowledge to local realities is beyond the scope of this paper and consists of targeted research that addresses very specific issues in the university environment, and for which there is institutional sponsorship

[9] HKUST was created in 1992 and was listed as number 40 worldwide in the Times Educational Supplement ranking of this year.
[10] Samir Amin. Delinking: towards a polycentric world, 1985.

both internal and external. Internal sponsorship is generally through seed grants and other forms of university funding[11].

The second form of scholarship, which is as important if not more important than localizing research, is the scholarship of teaching and learning. It is probably more important because teaching is the core mission of most universities in the region. Therefore research that supports this core mission is obviously the most important form of research. This basic equation of relating research to teaching seems to have escaped most universities where the scholarship of teaching and learning has barely surfaced[12]. This is largely due to the uncritical borrowing of university systems without due adaptation to local realities. A compounding factor is that local universities look up to elite institutions in the West to provide models of organization and in so doing tend to overlook their own idiosyncratic realities. Indeed, in research universities where the basic mission is to produce marketable research, it is logical that the latter fulfils the requirement of the knowledge market. However, when such a market becomes remote and difficult to access, attempts to mimic research endeavours in the research universities are bound to deliver second rate research. This has been fuelled by the proliferation of publication outlets that cater to this type of research. As a result, the research process and evaluation have become essentially 'bean-counting' mechanisms and meeting nominal criteria of promotion and other types of extrinsic rewards.

The paucity of such research is most visible in its limited impact measured first by the impact factor of the journals where it is published and secondly by the limited, if at all, citation of such research work. Being allegedly targeted at competing within the global research arena and having no impact, hence mustering very limited readership, this type of research amounts to not much at all and hardly benefits the institution where academics work. On the contrary, it takes away from the time that faculty could have devoted to other more relevant endeavours including teaching and other more relevant forms of scholarship.

This *modus operandi* is fostered by evaluation systems that put pressure on faculty to 'make the cut' in their research work, at the risk of having their work contracts discontinued. It is important to mention, however, that the whole process is not engaged on deceitful bases as might transpire from the elaboration above. On the contrary, be it universities or the faculty themselves, they genuinely invest money and effort respectively in research endeavours in the absence of an alternative game in town. Indeed, the alternative of doing irrelevant research is believed to be no research at all, which almost everybody is not willing to engage in because of its cost; for universities it amounts to a loss of academic capital hence a perception of being second rate, and for faculty it translates into the erosion of their marketability and a status of second class citizen in the university. What is essential then is to re-orientate the invaluable research efforts of faculty and the accompanying resources provided by universities to supporting the core mission of the university, namely teaching. We further elaborate below on the basic tenets of the scholarship of teaching.

4 THE SCHOLARSHIP OF TEACHING AND LEARNING

The concept of scholarship of teaching and learning was first introduced by Boyer (1990) and his colleagues at the Carnegie Foundation for the Advancement of Teaching when he defined an

[11] Many universities in the Gulf region run similar schemes. A nationwide scheme of research funding of this type is run in Tunisia by the Ministry of Technology and Scientific Research which is funding a large number of research labs and research units all over the country with research thematics being necessary local. A government agency, CNEAR, evaluates these research cells scattered all over the university system and in light of concrete outcomes, decides to either continue or discontinue funding them. Similarly, the recently launched initiative by Abu Dhabi University to develop a business case repository of local practices fits within this framework of developing a local research agenda, independent of international dynamics.

[12] Annual self-reports at the American University of Sharjah include a section on "contribution to learning and pedagogical research" along with 'discipline-based research.' However it was never made explicit whether such type of research is counted towards contract renewal and promotion decisions beyond the annual report-based merit increase.

extended representation of scholarship in academic practice. Accordingly, conventional research or the scholarship of discovery is only one form of scholarship, along with three other forms, namely the scholarships of integration (e.g., writing textbooks), application (service), and teaching and learning. His typology of scholarship is based on the simple premise that scholarship within a university setting has to address not only discipline-related problems and issues but also the processes through which discipline-based knowledge is produced. The scholarship of teaching and learning hence involves systematic study of teaching and/or learning and the public sharing and review of such work through presentations or publications (Kreber, 2003).

The narrow focus on research productivity that universities traditionally pursued has led to unbalanced careers in acadème (Huber, 2001) going against the professional inclinations of faculty to engage on more than the research front. The over-emphasis on research at the expense of other forms of scholarship endangered the core mission of universities which were increasingly called upon to fulfil different obligations towards a variety of stakeholders, chief amongst these the teaching obligations towards students. Conceived outside of the framework of scholarship, teaching has been evolving haphazardly, that is without considerations of pedagogy and subjected to the arbitrariness and personal philosophies of individual faculty members. Mastery of subject content is generally assumed to suffice for a successful transfer of knowledge. Empirical evidence suggests that over-emphasis pedagogies are more effective (Samuelowicz & Bain, 2001).

Policy agendas and changes at the supranational, national, and institutional levels have brought increasing pressures on academics to adjust their teaching practices in line with the requirements of the knowledge society for a over-emphasis pedagogy which substitutes the student as an active participant in the learning process to the meek recipient of teacher's knowledge (cf. Norton et al., 2005). However, faculty development programmes, meant to bring faculty in the disciplines to comprehend the new pedagogies are generally resisted because they are resented as external interference in the inner business of the disciplines. Moreover, pedagogical training is thought to be driven by epistemological assumptions of trainers, which is discipline-driven as well (generally by educationalists).

All of the above had made a discipline-specific scholarship of teaching a necessity. This would include discipline-specific pedagogies or disciplines as pedagogies (Mills & Huber, 2005), curriculum development, faculty development and evaluation, e-learning, departmental organization of teaching and learning, management of educational change, etc., and in general every process and activity that impacts the knowledge delivery process in a particular discipline. Educational research that normally aims at studying such processes of knowledge production is not undertaken within the disciplines so it is generally not adopted and has had little impact and relevance.

"Education's marginal status affects its ability to influence both academic theory and practice, making academics less likely to credit value of reading and writing about teaching and learning… 'discipline-specific pedagogy' comes with a set of often tacit assumptions about the appropriate subjectivity to be adopted by a scholar of the discipline" (Mills & Huber, 2005, p. 14).

It is indeed perplexing that disciplines perform research about the production delivery of many other activities and services in view of improving them whilst at the same time overlooking their own organization in doing such research. Mills and Huber (2005) refer to this as "a disinclination amongst most established disciplines to critically examine the conditions of their own production and reproduction." The reason for such an anachronism is historical and has to do with the discipline-specific conception of teaching and research whereby knowledge is the output of a loner's effort in examining and studying phenomena, basically the 'ivory tower' model. This conception of knowledge production could be termed mode 1 as opposed to mode 2 (Gibbons et al., 1994). In mode 2, knowledge production is driven by specific requirements such as quality assurance criteria and benchmarks whereas mode 1 is more epistemological with learning driven by a 'pursuit of the truth' in the researcher's idiosyncratic way. The discipline-specific as the cradle of knowledge production in mode 1 is being gradually overtaken by projects and structures of interdisciplinary in mode 2 as knowledge becomes increasingly discipline-specific and produced in various areas closer to its application (classroom).

The core element of this new form of scholarship is that of teaching and learning, which seeks to build a commons for the teaching profession (Huber, 2001). This commons however is not only discipline-specific but encompasses trading zones (Mills & Huber, 2004) wherein synergies between different disciplines are harnessed to improve knowledge production and reinforce a common identity across the academic profession. This is a stark departure from the current situation where there hardly any deals are made in the trading zone. It is indeed very rare for faculties of engineering to draw from the pedagogies and practices of faculties in the arts and sciences in their mode of knowledge delivery and vice versa. Teaching and learning are perceived to be driven by the specific requirements of the discipline, yet these requirements are never made explicit, let alone based on scientific methods of inquiry. This creates problems and resistance to change when new conceptions of teaching and learning become essential to upgrade the quality of knowledge.

Constructivism, which has been long advocated as the new pedagogy befitting the knowledge society (Samuelowicz & Bain, 2001) is relentlessly resisted because it is perceived as undermining the authority of the teacher and subjecting the absolute truth of knowledge to the whims of inexperienced learners. It is only through the development of a scholarship of teaching that teachers can grasp and appreciate the added value of constructivism and other concepts of teaching and learning for that matter. Matters are made worse when entire curricula and their delivery mode are shifted to align with new teaching paradigms. They are invariably resisted and misunderstood by a faculty body that grew accustomed to individual heuristics in their teaching delivery process.

The professionalization of teaching, which has been referred to by many as the devaluing of teaching, is looked at as an external interference by auditors into the inner working of a fundamentally 'artsy' profession. In their resistance to a process of normalization of teaching and the transparent sharing of experiences through the commons, some went as far as decrying attempts at fostering the nakedness of the teaching process under the disguise of quality assurance and professionalization (Shore & Wright, 1999).

Under such conditions, one might doubt that a scholarship of teaching even within the boundaries of the disciplines might be more successful than classical faculty development programmes and trans-disciplinary educational research. However one critical element has to be considered in the equation, namely disciplinary identities which act as normative cultures in defining the stand of faculty towards their environment (Becher & Trowler, 2001). Given that a scholarship of teaching is well within the bounds of the discipline, driven by native epistemologies, and serving as a balancing mechanism for the involvement of all faculty in some form of scholarship thus leading to the accumulation of research capital, its acceptability within the disciplines as an internal process is very likely. This will be doubly beneficial because on one hand it will bring faculty problem-solving skills to bear on teaching and learning and on the other hand it will lessen their resistance to constructivist pedagogies once they understand them and apply them within a scholarship framework rather than as administrative edicts.

In the last section of this paper, we briefly analyse the situation of universities in the Arab and Gulf regions and examine the possibilities that the scholarship of teaching and learning has to offer.

5 ARAB UNIVERSITIES AND THE TEACHING IMPERATIVE

The HE sector has developed significantly in the Arab region in general and the Gulf region in particular especially over the last decade with the establishment of many private universities that were created alongside the more established national institutions[13]. As of 2003, there were a total of 233 universities in 21 Arab States, 156 public and 77 private or non-governmental. The first generation of universities was built to cater to the development needs of Arab countries as they

[13] University of Bahrain, UAE University, King Saud and King Fahd Universities, the University of Qatar, Kuwait University, and Sultan Qaboos University among others.

set their industrialization and economic development processes in motion. They were generally successful in providing for the quantitative needs of public service and private business but were less so in terms of other critical variables:

"Among the ailments that various observers and reports have diagnosed the Arab universities with, one can mention declining quality, weak production of knowledge, absence of independent vision, inadequate curricula, and centralized governance." (Guessoum & Sahraoui, 2005)

Though more than 100,000 faculty members are involved in teaching and research in institutions of HE in the Arab World with an equally large number of Ph.D. holders, research production is thought to be limited:

"Without research centers and laboratories, and without a critical mass of post-graduate programs and students, and proper budgets, there would be little renewal or production of knowledge, but merely transmission of what is already known. It is imperative that universities in the Arab States develop programs of postgraduate study, centers and laboratories for research, to provide students and faculty alike with the opportunity to produce new knowledge to the benefit of their societies." (Bashshur, 2004)

The massive expansion of the system and the subsequent creation of poorly funded private universities meant that the focus was on teaching rather than research and scholarship. In some Gulf countries, however, the new private universities like the American University of Sharjah, and institutions implanted within Qatar Education City, just to give a few examples, emphasized research alongside teaching. Yet again, as most teaching happens at the undergraduate level and disproportionately in the humanities and social sciences rather than in science and technology fields (World Bank, 2002), little 'marketable' knowledge has been produced by Arab universities and highly qualified faculty resources are being consumed by the heavy load of the teaching process.

The majority of students in the Arab world still receive their Ph.D. from Western universities, reflecting the lack of a research environment amenable to nurture the development of a track record of research graduates at the highest levels of research qualification[14]. However the focus on teaching has not been accompanied with an improvement in quality. To the contrary, as Bashshur (2004) points out:

"Style of teaching is another area reflecting quality. Here, there is universal agreement that the dominant style is the lecture, and the emphasis is on memorization. "The lecture", in UAE University, "is delivered in a rhetorical manner, while it is in fact not more than a summary of what is in the book..." The professor refuses to answer questions other than repeat what he has already said; discussion or dialogue is not known, and students are expected to reproduce what they have heard; otherwise their grades will be low... Observations and impressions on teaching in Arab universities abound, and they come from different sources and countries. Rarely one reads about or sees situations involving dialogue or experimentation, let alone what is described in the new pedagogical literature as "interactive" teaching." (p. 80)

Even quality assurance processes that have come to bear on many Arab universities as part of the accreditation processes have not changed much to this reality as institutions learn how to cope with the formal requirements of QA without really altering their teaching processes for more effectiveness. The problem lies not in the unwillingness to change amongst either faculty or universities but in the absence of a body of scholarship to indicate the way of engaging fundamental change in teaching thus benefiting from the commons that is being developed worldwide by the community of the scholarship of teaching and learning. It is even unlikely that the concept and the related initiative were heard of.

Nonetheless and as stated earlier, a great opportunity to develop such a scholarship is awaiting fulfilment if only priorities and resources were redirected accordingly. The critical mass of Ph.D. holders who are disconnected from the knowledge economy because of the sheer load of teaching

[14] "The majority of Saudi students working on their Ph.D. programs still do this outside the country (more than 90% of them). In Jordan Ph.D. programs are offered in two universities only of the six public ones; none of the private universities offers a program of study beyond the Bachelor." (Bashshur, 2004).

could thrive in the three forms of scholarships other than that of discovery. This would have great and positive repercussions not only on the quality of education provided, but also on the careers of faculty and the student experience. However, the development of a scholarship of teaching and learning cannot just happen as a result of the interest of few academics. Institutional changes are required and include the following:

- Provision of seed grants and other types of funding for the scholarship of teaching and learning;
- Legitimation of the scholarship of teaching and learning within the disciplines;
- Revision of performance evaluation for promotion and career advancement including contract renewal in ways that values this type of scholarship. Current methods of retaining only publication outlets within the nominal area of the candidate and minimizing the rewards for writing textbooks for instance have not encouraged faculty interest in teaching-related scholarship;
- Adoption of teaching evaluation instruments and methods that emphasize scholarship-driven delivery rather than the traditional transfer of knowledge. This can prove problematic because usually students express satisfaction with teaching that allows them to 'make the grade' (Becker et al., 1968) rather than one that benefits them more but puts a heavier burden on them to become active learners;
- Development and dissemination of a teaching commons first within the university and secondly outside of it and especially at a regional level. Indeed and as with the scholarship of discovery, there is no guarantee that findings will be implemented in the classroom unless proper dissemination is planned. Faculty and teaching development workshops and seminars should serve as conduits for the exchange and transfer of successful experiences;
- The professionalization of teaching and learning. Caution has to be exerted not to centralize such function by imposing the disciplinary epistemology of one area, be it education, on the rest of the disciplines. Indeed, if a faculty development centre is to be created, its role should be to provide resources for the scholarship of teaching and learning to take hold within the disciplines and departments.

6 CONCLUSIONS

The scholarship of teaching and learning has been advocated above as providing opportunities for academics in Arab universities to pursue one form of scholarship where they would not be second rate. Moreover, universities in the region who have opted for teaching rather than research as a core mission can benefit by improving the teaching process and delivering a high quality service to students and their employers. We have to stress in the conclusion that the scholarship of teaching and learning is not sub-standard or suitable only for those who cannot perform well in the scholarship of discovery. On the contrary, it will generally be of higher quality because it is more relevant to the university environment, hence creating a level playing field between researchers in different learning environments. This is not the case for the scholarship of discovery where research capital is concentrated in fewer locations worldwide. The Shanghai University ranking which is completely based on the scholarship of discovery illustrates such disparities with a concentration of top ranked universities in the USA, UK, and a few other industrialized countries. Moreover, the scholarship of teaching and learning is not being advocated as a replacement for that of discovery whenever the latter is of good quality, and such instances exist throughout the region; rather it is an additional and important form of scholarship that is within reach of a larger number of people. Similarly and though this was not covered in this paper, two other forms of scholarship, namely of application (service) and integration (textbooks) are equally important. Along with the scholarship of teaching and learning, they will help strengthen universities and better prepare them in their knowledge production mission. Finally, such scholarship will help create an institutional memory for HE in the region and put an end to the process of uncritical policy borrowing in the implementation of educational changes including appropriate pedagogy (Johnson, 2006).

REFERENCES

Barnett, R. 2006. *Reshaping the University: New Relationships Between Research, Scholarship and Teaching.* Milton Keynes, UK: SRHE and Open University Press.

Bashshur, M. 2004. *Higher Education in the Arab States.* Beirut: UNESCO Regional Bureau for Arab States.

Becher, T. & Trowler, P. 2001. *Academic Tribes and Territories: Intellectual Enquiry and the Cultures of Disciplines*, second edition. Milton Keynes, UK: SRHE and Open University Press.

Bourdieu, P. 1988. *Homo Academicus.* Cambridge: Polity Press.

Boyer, E. 1990. *Scholarship Reconsidered: The Priority of the Professioriate.* Princeton, NJ: Carnegie Foundation for the Advancement of Teaching.

Deem, R. 2006. Conceptions of Contemporary European Universities: To do research or not to do research? *European Journal of Education*, 41(2): 281–304.

Guessoum, N. & Sahraoui, S. 2005. *The Role and Impact of American Universities in the Arab World.* Presented at the AGHA KHAN CONFERENCE on "Educational Systems in the Muslim World", London, February, 2005.

Gibbons, M., Limoges, C., Nowotny, H., Schwartzman, S., Scott, P. & Trow, M. 1994. *The New Production of Knowledge: The Dynamics of Science and Research in Contemporary Societies.* London: Sage.

Henkel, M. 2000. Academic Identities and Policy Change in Higher Education. London: Jessica Kingsley.

Huber, M. 2001. Balancing acts: Designing careers around the scholarship of teaching. *Change*, July/August.

Johnson. 2006. Comparing the trajectories of educational change and policy transfer in developing countries. *Oxford Review of Education*, 32(5): 679–696.

King, R. 2004. *The University in the Global Age.* Basingstoke: Palgrave MacMillan.

Kreber, C. 2003. The scholarship of teaching: Conceptualizations of experts and regular academic staff. *Higher Education*, 46(1): 93–121.

Lucas, L. 2006. *The research game in academic life.* Milton Keynes, UK: SRHE and the Open University Press.

Mills, D. & Huber, M. 2005. Anthropology and the educational trading zone: Disciplinarity, pedagogy and professionalism. *Arts and Humanities in Higher Education*, 4(1): 5–28.

Naidoo, R. 2004. Repositioning higher education as a global commodity: Opportunities and challenges for future sociology of education work. *British Journal of Sociology of Education*, 24(2): 249–259.

Norton, L., Richardson, J., Hartley, J., Newstead, S. & MAYES, J., 2005. Teachers' beliefs and intentions concerning teaching in higher education. *Higher Education*, 50: 537–571.

Samuelowicz, K. & BAIN, J. 2001. Revisiting academics' beliefs about teaching and learning. *Higher Education*, 41: 299–325.

Slaughter, S. & Leslie, L. 1997. *Academic Capitalism: Politics, Policy and the Entrepreneurial University.* Baltimore: The Johns Hopkins University Press.

Sharp, S. 2005. The Research Assessment Exercises 1992–2001: patterns across time and subjects. *Studies in Higher Education*, 29(2).

Shore, C. & Wright, S. 1999. Audit culture and anthropology: Neo-liberalism in British higher education. *JRAI*, 5: 557–575.

World Bank, 2002. *Constructing Knowledge Societies: New Challenges for Tertiary Education.* Washington, DC: Education Group, Human Development Network.

Higher Education in the Twenty-First Century: Issues and Challenges – Al-Hawaj, Elali & Twizell (eds)
© 2008 Taylor & Francis Group, London, ISBN 978-0-415-48000-0

The role of technologically-enhanced and interactive e-learning systems in the delivery and content quality of degree and continuing education programmes in management

Deniz Saral
Chairman, Business and Management Programs, School of Business and Technology, Webster University, Geneva, Switzerland

ABSTRACT: "There is a growing crisis in management education that threatens to damage the quality of education and stifle the opportunity to achieve excellence for future students in business schools around the world. In simple terms, there are too few individuals working toward their doctorate in business to cover all of the predicted openings in the future. This shortage, if left unchecked, can severely damage the reputation and excellence of business schools around the globe."[1]

In August 2002, The AACSB International Management Education Task Force (METF) issued a landmark report, *Management Education at Risk*, which identified and prioritized the most pressing issues facing management education. One of the foremost METF concerns related to an emerging global doctoral faculty shortage in business. The AACSB International Board of Directors responded by creating the Doctoral Faculty Commission (DFC). The DFC came up with a series of recommendations that could alleviate the shortage of business doctorates in the future.

The author examines the recommendations contained in the DFC report under the light of recent developments in Europe and concludes that the most promising alternatives are: (a) Introduction or expansion of doctoral programmes for executives, and (b) Using technology-enhanced e-learning models to reduce the cost of doctoral education. The author then presents the latest efforts of the European Federation of Management Education (EFMD), and EFMD CEL (Certification of E-Learning), which is EFMD's certification arm for **teChnology-Enhanced Learning**. The fact that there are only a handful of programmes that have earned EFMD CEL status so far, indicates how demanding the standards in this area have become.

1. WHAT ARE THE CRITICAL FACTORS CAUSING THE CURRENT SHORTAGE OF PH.D. APPLICANTS IN BUSINESS SCHOOLS?

Throughout history, business schools had to compete with industry to attract talent to their Ph.D. programmes. In areas such as accounting and finance, the industry won often. In traditional areas, such as economics, the reverse was true. Since the late 1990 s, industry has been overly successful in diverting talent from business schools in all areas. This lack of ability to compete with industry

---

Clicks **or** *Bricks?* Is this the fundamental question for universities in the 21st century? If the word *or* is used in the inclusive sense, my answer is definitely yes. However, institutions of higher learning and training that insist on "predominantly bricks" will no longer be able to compete with those adopting mixed strategies. Virtual learning has already arrived.

---

[1] Beta Gamma Sigma International Exchange: Is there a doctorate in the house? Vol. 2, No. 2, 2004.

has become so acute recently that a good number of business schools had to terminate their Ph.D. programmes in the USA. The "Ph.D. halls are empty" message echoes equally on the corridors of traditional and state-supported universities in Europe. Hence the raison *d'être* for Management Education Task Force's (METF) landmark report in 2002 entitled **Management Education at Risk** and the subsequent recommendations by the Doctoral Faculty Commission (DFC)[2].

Various reasons have been advanced to explain why there is the shortage of Ph.D. candidates knocking on business school doors: Ph.D. programmes (1) take too long (3–5 years); (2) are not flexible (full-time attendance is required, which does not permit the candidate to work); (3) are too costly when the opportunity cost to the candidate is taken into consideration. This cost is incurred when the candidate chooses a Ph.D. programme instead of working in industry after an MBA degree; (4) are too theoretical and thus do not address the current needs of business; and (5) still require that knowledge is transferred from mentor to candidate in a low student-mentor ratio environment, which is a very costly process for universities. The culprit is most likely a combination of all factors proposed. However, the conclusion is uniform. These factors, and others, have reduced Ph.D. candidate supplies to critically low levels and business schools have to find alternatives to increase inflows into their doctoral programmes. Otherwise, the delivery and the quality of management education will be at risk.

2. WHAT ALTERNATIVES DO BUSINESS SCHOOLS HAVE?

Faced with high global demand from business for well-trained managerial talent, business schools enthusiastically increased their MBA programme capacities. Numerous traditional universities in Europe joined in and furiously tried to convert their economics departments into business schools. Their faculties had already realized how fast the demand for non-business programmes was eroding. This phenomenon resulted in a proliferation of MBA programmes across the globe, but, most significantly in the USA and Western Europe. Intense competition followed. Today, even the highest ranked business schools have to spend fortunes to recruit MBA students. They have to go far and expand offers across continents. Recently, INSEAD forged a strategic alliance in Saudi Arabia and the London Business School completed a parallel feat in the UAE. IMD (Lausanne, Switzerland), the other school in the top three in Europe, seems to resist temptation so far. This is largely due to IMD President Peter Lorrange's strategic model, which requires that "Mohammed must come to IMD in Lausanne"[3]. Thus, if a candidate wants to earn an IMD MBA, s/he must spend a full-time year in Lausanne. Their Executive MBA programme has a similar requirement, that is, all candidates must be ready to spend weeks at a time in Lausanne over two years.

How many business schools can afford IMD's strategy?

The attempt to convert traditionally-educated professors in Europe (and in the USA) to become business school teachers failed. I know a number of European state universities that failed to attract the minimum required enrolments in their MBA programmes in 2006 even though they are AACSB accredited. And, they had worked so hard to obtain their accreditations!

If business schools can't attract new talent in sufficient numbers to their Ph.D. programmes and are also unable to retrain their traditional faculty to become effective teachers in MBA or Ph.D. programmes, what alternatives do they have? In the rest of this paper, I will illustrate how two of the alternatives listed in the Doctoral Faculty Commission (DFC) report can be fused and brought to bear jointly to elevate the restricted supply of doctoral candidates in business schools.

---

[2] Both are AACSB International bodies.

[3] A few years ago, during an EFMD Annual Conference Workshop in Brussels, Peter Lorrange publicly questioned INSEAD's strategy to open a second campus in Singapore and since then has steadfastly defended IMD's strategy of not opening additional campuses anywhere in the world. Given his chronological age and his imminent retirement, however, the next IMD president may well change this strategy, which has worked admirably for IMD up to now.

3. INTRODUCTION OR EXPANSION OF DOCTORAL PROGRAMMES FOR EXECUTIVES

Can business schools recuperate the talent they earlier lost to industry and encourage them to undertake doctoral studies? The response is affirmative provided that the doctoral programme (1) is based on applied research, which allows the executive to effectively apply the experience and talent s/he has accumulated over the years; (2) is flexible, that is, the programme can be taken on a part-time basis; (3) is industry relevant so that the doctoral candidate can apply the results of his/her newly acquired research and knowledge readily, and (4) is able to motivate the doctorate to teach in business schools on a part-time basis after the completion of the degree programme.

Thierry Grange, the Director General of *Grenoble École de Management* (GEM) in France, describes this phenomenon as a process for "knowledge/talent recycling", which will enable business schools to recuperate the talents they lost to industry years ago. GEM is among a good number of highly recognized business schools and universities that offer a Doctorate in Business Administration (DBA) degree programme.

There is a potential shortcoming of this model, however. If the DBA candidates are not provided with an Integrated Learning System (ILS) through which they can (a) stay in constant contact with not only each other but with their mentors and (b) learn by interacting with the e-learning system imbedded in the ILS, then the DBA program may lose a good number of their DBA candidates. It is important to keep in focus that the business executive has been trained and motivated to work towards measurable, time-based, continuous and compensated objectives. They will therefore become "high-maintenance students". If the DBA programme administrators and mentors are not equipped to respond to the executive's individual needs and requirements, and this is very costly indeed, then they must invest in an ILS that will respond to the executive's needs for training in academic research methodology and interactive learning. Intermittently scheduled 4- to 5-day "anchoring workshops", which bring all of the DBA candidates and the mentors together, will complete the DBA training process.

4. WHAT ARE THE CHARACTERISTICS OF AN INTEGRATED LEARNING SYSTEM?

The ILS environment combines the following features:

- Fully interactive web-based courses,
- Face-to-face "anchoring workshops" with real-life cases, examples and studies,
- Virtual classroom sessions bringing participants together with mentors,
- Evaluation tests designed to measure the effectiveness of learning,
- Simulations modelled to reflect real-life applications of the knowledge gained,
- Project work allowing participants to put theory into practice,
- Access to research databases and information to enable the participants to conduct research,
- Mentor support for all of the aspects listed above.

Designing and successfully operating an ILS requires accumulated know-how and specialized talent. An ILS, therefore, is not a panacea. It has been demonstrated to be highly effective, however, and one which costs less than traditional doctorate and EMBA programmes to operate in the long run.

5. HOW IS THE QUALITY OF CONTENT AND DELIVERY EFFECTIVENESS OF AN *ILS* MEASURED?

Since the quality of both products and programmes varies widely, there is a cause for concern. In order to address this, EFMD formed a joint initiative with the *Swiss Centre for Innovations in*

Learning (SCIL) at the University of St. Gallen. "There is a lot of very poor e-learning around and the introduction of accreditation that gives a quality signal is a very important thing to be doing," said Mark Fenton-O'Creevy, director of programmes and curriculum at the Open University Business School[4]. The fundamental objective of the CEL process is to raise the standards of e-learning programmes worldwide.[5]

How much time and effort will it take a business school to develop an ILS of its own? Unfortunately, the answer to this question is not very encouraging, because the design, testing, development, constant maintenance and upgrading of an ILS make up a complex process to manage. Many e-learning products have crashed and burnt because their developers based them mostly on technology rather than teaching and learning. Therefore, learning how to design an ILS requires a long transition period and is costly.

There are nevertheless a handful of organizations that have already mastered the ILS. Some are also certified by the EFMD CEL and make excellent candidates for forming strategic alliances. The responsibilities are then shared between the parties. The local/regional university concentrates on (a) recruiting DBA or EMBA candidates, (b) designing "anchoring workshops" to be executed in central locations so that candidates from afar need not travel long distances and (c) training its own faculty to become DBA or EMBA mentors within the ILS. On the other hand, the institution that owns the ILS (with EFMD CEL status) applies it towards the training/learning objectives of the DBA or the EMBA candidates.

During its 2007 Annual Meeting in Brussels, June 10–12, EFMD CEL will award this prestigious status to Kavrakoglu Management Institute's (KMI) Executive MBA program.[6] Ibrahim Kavrakoglu, KMI's director and mentor, earlier reported that it took them over USD 60 million (BHD 22.5 million) of investment and nearly ten years to develop and verify their ILS! He is convinced that there is a very strong place for virtual learning in the 21st century. So am I.

---

[4] Financial Times, www.ft.com, 21.03.2005.

[5] Visit www.efmd.org

[6] www.kavrakoglu.com

Higher Education in the Twenty-First Century: Issues and Challenges – Al-Hawaj, Elali & Twizell (eds)
© 2008 Taylor & Francis Group, London, ISBN 978-0-415-48000-0

Ingredients of successful partnerships among higher education institutions: The case of the University of Westminster

Kadom J.A. Shubber
Higher Education Academy, UK, and Westminster Business School, University of Westminster, UK

ABSTRACT: This paper looks at the topical issue of collaboration agreements among higher-education (HE) institutions, considering various aspects of these arrangements, as well as the ingredients which deliver success in partnerships. Three main phases are probes, namely initial development-*cum*-contract formulation, day-to-day management, and periodic re-validation.

In this regard, the experience of the University of Westminster is looked at, in order to identify different types of partnership agreements, as well as assess the significance of effective partnerships in contributing to the attainment of overall institutional objectives. Whether we consider levels of enrolment, research output, gross revenues, financial surpluses, or institutional image, successful partnerships can act as a major strategic dimension of overall activity, particularly where respective institutions complement one another in certain major features.

The paper classifies pertinent synergies into three groups. The first relates to the focus on theory or practice; the second relates to academic guidance; the third concerns relative advantage in enrolment. The paper also contains some data regarding current partnership agreements between Westminster and other HE institutions.

1 INTRODUCTION

It has probably become a prime observation of both academics and students that partnerships or collaboration agreements among universities (and other institutions of higher education) have become a hallmark of this generation. As business activities have become increasingly globalized, higher education institutions have been following suit, attempting in the process to accomplish their objectives of survival, healthy finances, and enhanced image.

This glaring internationalization of higher education has several facets that are inter-twining and inter-relating. Among these facets are the flow of students across borders, the increasing number of distance-learning schemes of study, joint ventures among universities, and co-operative programmes of research.

A further aspect of the globalization of higher education has been the steady growth of collaboration/partnership arrangements among universities and other higher education institutions. These arrangements have extended to exchanging students, the awarding of degrees, the award of a franchise (franchising), synchronization of study programmes, supervision of research students, and exchange of academic staff.

Some evidence exists that co-operation in these areas has assisted with student recruitment, increasing revenues, and achieving higher financial surpluses (or reduction of deficits). In addition, this type of co-operation has probably contributed to up-grading the image of respective institutions, and providing more choice to students and academics, as well as building a more open ambience that is responsive to ideas and conducive to healthy dialogue.

2 THE APPROACH AT WESTMINSTER

The University of Westminster espouses a practical and positive approach towards collaboration with other HE institutions, in that it is prepared to consider any worthwhile propositions for partnership, whenever clear benefits to both parties can be envisaged, and where the ingredients of success are present. Currently (May 2007), some 1,680 students of the University are covered by collaboration agreements. This makes up almost 8 percent of the overall student body of around 22,000.

At present, the University has a total of 59 agreements with collaborative partners, and the respective agreements cover 44 courses (both under- and post-graduate). As Table 1 shows, partnership agreements with UK-based institutions represent around 56 percent of the grand total. This is followed by agreements with institutions based in Western Europe (18.6 percent), then in joint third place institutions based in East Asia and CIS (Commonwealth of Independent States), with 12 percent each.

Table 2 presents a breakdown of the categories of collaboration between the University and its partners. External validation tops these categories with just over two-thirds of the total, followed by dual awards with almost 19 percent, then franchising.

3 REQUISITE CONDITIONS

It would be wrong, however, to imagine that collaboration agreements are a panacea, which can put to rest all problems and difficulties we face at higher education institutions. Of equal importance is the need to recognize the right conditions for these arrangements to be workable and beneficial to all parties. Several issues arise in this regard, including the exact type of partnership to be adopted, proper formulation of the co-operation agreement, day-to-day management of the arrangement, trouble-shooting mechanisms, and periodic re-appraisal and re-validation of the collaboration scheme.

Table 1. University of Westminster: Regional Breakdown of Collaborative HE Partners.

Region or country of partners	Number of partners	% of total
UK	33	55.9
Western Europe	11	18.6
East Asia	7	11.9
Commonwealth of Independent States	7	11.9
Others	1	1.7
Total	59	100.0

Source: Academic Registrar's Department, University of Westminster, U.K., May 2007.

Table 2. University of Westminster: Categories of Partnership Agreements with HE Institutions.

Category of agreement	Number of agreements	% of total
External Validation	39	66.1
Franchising	8	13.6
Dual Award	11	18.6
Distance/Flexible Learning	1	1.7
Total	59	100.0

Source: Academic Registrar Department, University of Westminster, U.K., May 2007.

Perhaps, the essence of any such scheme is that the co-operating institutions must complement one another. That is, there has to be some synergy between them, so that each must enjoy a relative advantage *vis-à-vis* the other.

Three main classes of synergies may be mentioned in this regard, whereby one or more of these might be embodied in any one situation.

3.1 *Academic versus empirical focus*

One institution may be more focussed on theory, while another is more in tune with the practical world. Students transferring between two such institutions in the context of a specific course of study would clearly benefit from the move. This has been the case with the *Master of Science in International Finance* (MSc IF) course that the author has been leading since September 2001[1]. Students at Westminster can spend the first semester at Westminster where they take up the more fundamental modules, with an option to move to *CERAM Sophia Antipolis* in the second semester. The same avenue is available to CERAM students taking a course with the same title.

Most academic staff teaching on the course at CERAM have a part-time commitment within the financial services sector, while most of their counterparts at Westminster are full-time academics. Some class sizes at Westminster are relatively large, with students from other courses taking the same modules. CERAM class sizes for the same course have been kept fairly small, thereby ensuring comparatively intimate monitoring of students' attendance, progression, etc. In addition, an internship is a core element of the MSc IF course at CERAM, which requires the student to undertake work within finance at a relevant organization for several weeks[2].

MSc IF students who go through the exchange successfully and pass all the modules will obtain two degrees (with the same title) from Westminster and CERAM. This is despite the fact that the two institutions do not have an identical curriculum for this course. The number of modules taken by the students, as well as module titles and contents, are somewhat different between the two institutions. For grading purposes, the weights of the respective modules also differ.

Yet, looking at the two programmes in an holistic fashion, it is possible to argue that students going through the course at either institution (or those benefiting from the exchange) will actually take in a similar totality of knowledge/experience. When they go through the exchange, they take advantage of the different cultures/structures of the two institutions, and can recognize that the learning process at each institution complements that at the other. This is in addition to obtaining two degrees.

3.2 *Academic guidance and sanctioning*

A further avenue for co-operation is where one institution enjoys a higher academic profile and/or image, and hence be in a position to benefit the other in this regard. Alternatively, it can be said that one institution enjoys relative prestige, due to being placed in a higher league within a certain specialist area, and hence may advantageously confer an award, design curricula, or oversee the quality of delivery.

The higher-league institution can thus help with curriculum design, monitoring delivery of programmes, ensuring proper assessment of students, and quality-assured award of degrees. In addition, the senior partner may assist with the provision of academic staff, as well as admitting respective graduates from the junior institutions for further study.

---

[1] Over the period 2001 to 2006, the author was the course leader of the MSc in International Finance (MSc IF) at Westminster Business School. Full-time Westminster students on the course could spend the second semester at CERAM Sophia-Antipolis in southern France. During the current academic year (2006–2007), the author is the joint course leader of the programme. CERAM students on a course of the same title could likewise spend their second semester at Westminster.

[2] It is noteworthy that CERAM Sophia Antipolis is a business school set up by the employers' federation in the southern region of France. Hence, the emphasis on the practical aspects of business education.

Looking at the experience of Westminster, this kind of co-operation encompasses both "external validation" and "franchising". Taking these two groups together, we notice that they make up almost 80 percent of the total number of partnerships.

3.3 *Enrolment*

Last, but not least, in importance is the issue of student enrolment. One institution may be better placed than the other to procure enrolment of new students. Alternatively, each institution may possess a relative advantage in this area with regard to certain types of students, in terms of their ethnic background, nationality, region of domicile, etc.

A local HE institution in, say, the Gulf region, or in China, may possess an advantage, in terms of the capability to recruit students within the region, *vis-à-vis* another HE institution in Europe. Advice, expertise and/or personnel may be provided by the European institution within the context of a co-operation scheme between the two.

While human mobility has shown an unmistakable upward trend over recent decades, the onset of the 21st century has brought with it some sinister and worrying events, which have caused a re-think by leaders in the arenas of both politics and domestic security. One clear and noticeable consequence of this has been the tightening of the conditions for granting travel visas by European and North American governments.

Thus, US universities suffered a sharp drop in recruitment of foreign students in the years following the terrorist events of 11 September 2001. A not-too-dissimilar phenomenon has been evidenced by UK-based HE institutions after the explosive attacks of 7 July 2005 in London.

All this has convinced European and North American leaders within the HE sector that one possible avenue is to design schemes and programmes which allow prospective students to remain in their country/region, while allowing them to enrol on programmes/schemes of study that comply with the quality standards sought.

4 MAIN PHASES

The starting point for any partnership arrangement is critical. Ideas for collaboration may arise at conferences and/or seminars, or through personal contacts. Other possible starting points are intermediaries, consultants, advertisements, or initiatives taken by one institution to seek a suitable partner.

Whatever the trigger, any proposed collaboration agreement needs to be probed in a detailed fashion, so that the rights and obligations of either party are defined without any ambiguity. Each party will need to see clear benefits from the arrangement, and these must be viewed as substantial enough to outweigh any costs or commitments.

The product of this first phase—if successful—is the formulation of a written agreement, which spells out in sufficient detail the terms of the co-operation. Mutual visits by academics/administrators to the premises/facilities of the other institution, group/individual meetings, and ample discussions over several issues will be integral parts of the negotiations leading to the birth of the collaboration contract.

Implementation and day-to-day management represent the second main phase. Issues may arise requiring some adjustment/re-orientation, in order to oil the wheels of proper and sound management. There is always the possibility that liaison officers, course directors, and others in managerial positions will face matters not envisaged in the written paragraphs.

In the case of the Westminster/CERAM exchange relating to MSc IF, CERAM adopted the practice of vetting applicants wishing to cross the Channel northwards, in order not to exceed the agreed usual ceiling for the exchange, i.e., 10 students *per annum*.

In the case of Westminster, no such restriction needed to be put in place, as the number of those wanting to go to France was comparatively limited.

Perhaps, the attractions of a major cosmopolitan city have always been a major advantage to Westminster, in addition to the location of the university in the heart of the UK capital. This is despite the excellent facilities that CERAM provided, its healthy environment within the French Rivera, friendly ambience, and small class sizes that CERAM adhered to.

Table 3. Record of MSc IF exchange students CERAM and University of Westminster.

Academic session	From U of W to CERAM	From CERAM to U of W
2005/06	6	5
2004/05	3	9
2003/04	10	9
2002/03	7	6
2001/02	10	8
Average	7.2	7.4

Source: MSc IF Course Leader's Monitoring Report Academic Session 2005/06, Westminster Business School.

To cope with this issue, Westminster adopted the practice of inviting a small delegation from CERAM to meet each incoming MSc IF cohort during the induction period. The CERAM delegation would talk (and provide visual back-drop) to the new students about the study environment at CERAM, in addition to matters relating to accommodation, library/computer facilities, curriculum, etc.

All in all, it has been possible to keep a fairly equivalent two-way traffic between the two institutions, as evidenced by Table 3. It is noteworthy here that the Westminster/CERAM collaboration agreement stipulates that no fees will need to be paid by either institution to the other, as long as the annual number dispatched by that institution is ten or lower.

It is thus clear that several types of issues may require attention once the agreement enters the implementation phase. Apart from the selection of students to be exchanged, there might be complaints over academic delivery, synchronization of grades, examination arrangements, length/timing of semesters, assessment of students' work, and supervision of student projects.

It is always preferable if specific provisions in the agreement can be resorted to, in order to resolve these problems. However, in many cases there is no substitute for direct dialogue among those concerned, so as to find an amicable solution. In some cases, supplementary agreements (e.g., over grading systems) may have to be negotiated and agreed.

Finally, there is the phase of re-validation. Re-validation is undertaken periodically every few years. At Westminster, the maximum period granted to any course on re-validation is six years, but frequently the re-validation panel grants a period lower than this.

The re-validation process provides a crucial opportunity to consider all aspects of the course, including the curriculum, level of enrolment, job destinations of graduates, views/appraisals put forward by students, and pertinent assessment by academics from other universities and prospective employers. Any collaboration arrangement attached to the course will also be investigated during this process, in order to evaluate how it is functioning, and hence it is imperative that participants from both parties make active contributions to this process.

It is often the case that the outcome of re-validation recommends certain changes in the collaboration arrangement. These changes may pertain to the convergence of the curricula of the two institutions, re-alignment of their grading systems, provision of extra information/facilities to students, visits/exchange of academic and/or administrative staff, and sharing of relevant reports/information.

Higher Education in the Twenty-First Century: Issues and Challenges – Al-Hawaj, Elali & Twizell (eds)
© 2008 Taylor & Francis Group, London, ISBN 978-0-415-48000-0

Higher education mapping of GCC countries: An analytical framework of strengths and opportunities

Purnendu Tripathi & Siran Mukerji
Arab Open University, Riyadh, Kingdom of Saudi Arabia

ABSTRACT: Higher education with international quality parameters has been one of the most important aspects of discussion for policy makers in the field of higher education of this region. There are 62 universities in these countries with highest number, 21, in the Kingdom of Saudi Arabia, followed by the UAE (15) and at least four in Qatar. Hence the GCC countries together present an excellent opportunity for growth and development of higher education.

The GCC countries also provide an excellent platform for area-specific development of educational programmes and strategic partnership with international agencies involved in quality assurance in higher education. Looking at the need of the future generation and the enormous potential for growth in economy, it is essential that an analytical study of universities of GCC countries should be undertaken so that a networked collaborative framework could be developed amongst these countries for making this region a centre of excellence in higher education.

In the light of an analytical approach as stated above, the present paper identifies the following objectives for brainstorming:

a. To examine the potential of universities in GCC countries on the various parameters of academic excellence;
b. To identify the areas of higher education where the possibility of growth and development is still untapped;
c. To design and develop a strategic partnership framework for the universities of the GCC countries and developing this region as a centre of excellence in higher education across the world;
d. To develop a policy document entitled "GCC Mapping of Higher Education" which can suggest a set of strategies to face the challenges in the field of higher education in 21st century.

1 INTRODUCTION

Education has been one of the most important dimensions of development for all the countries in view of its contribution towards national economic and social growth and development. Universalization of education has been focussed in a number of conferences all around the globe such as *The World Conference on Education for All* held in Thailand in 1990, in which governments pledged to achieve education for all by the year 2000. Subsequently the progress was reviewed in 1996 in Jordan. This was followed by the World Education Forum in Senegal in 2000, in which the *Dakar Framework for Action* was developed which set educational goals for the international community.

According to this framework, education was identified as a human right and endured commitment towards education for all from different countries. This paved the way for the formulation of Millennium Development Goals (MDGs) which was passed by the UN General Assembly in a special session in the fall of 2000. This further strengthened the commitment of all the countries towards education for all. As a part of MDGs, education was one of the indicators for which specific time goals were formulated. Apart from the goal of achieving universal primary education, it focussed on tertiary education, and gender equality and the empowerment of women through tertiary education preferably by 2005 and at the latest by 2015.

As a part of the commitment towards development through education and fulfilment of the international goals, the GCC countries have been focussing on a comprehensive formulation of educational plans through a number of resolutions starting with the deliberations of the Seventh Session of the Supreme Council held in Muscat in December 1985 to those of the Twenty-fourth Session in Kuwait in December 2003. These resolutions have been emphasizing education and its importance in these countries. As is evident, education has continued to be one of the key issues of concern for the GCC since its inception. The Charter of the Council, the Economic Agreement and the Strategy of Comprehensive Development has focussed on education which shows the importance accorded to education by the Council. Having realized the challenges faced by these countries towards extending higher education to all segments of the society, these countries have initiated concrete measures in this direction by establishing a number of universities, providing various higher educational programmes, collaborating with international educational institutions of repute and conducting continuous assessment of progress in this direction. What follows is an illustration of the present scenario of higher education in this region.

2 HIGHER EDUCATION: THE PRESENT SCENARIO

Over the years, GCC countries have witnessed tremendous growth in education sector especially in higher education. In this section, we will analyse the potential of higher education in GCC countries on various dimensions affecting the growth and development of higher education. These dimensions are Adult and Youth Literacy, Infrastructure set up for higher education such as a universities' network in each of the GCC countries, and the availability of trained and qualified personnel. The dimensions also include enrolment strength in the various universities and international mobility of students (i.e., how many students were sent abroad for higher education and how many foreign students joined the universities of the GCC countries), and quality education streams. These dimensions will help us to initiate composite analysis of higher education and subsequently will help us in assessing the strengths and weaknesses of higher education system and also in formulating the GCC higher education mapping.

3 ADULT AND YOUTH LITERACY

The first dimension selected for the analysis of higher education is adult and youth literacy. The dimension was selected because it gives us possible potential figures of people who will take up higher education in the coming years. The data presented in Table 1 give some interesting revelations.

From Table 1, it is evident that data are not available for the UAE; however, the literacy rate in Adult and Youth seems to be impressive in almost all the countries, the highest being 93% in Kuwait. Similarly, male and female literacy is also highest in Kuwait followed by Qatar, Bahrain, Oman and Saudi Arabia. In illiteracy in women, Qatar again shows an impressive figure with the least percentage of women as illiterate (29%) while the highest percentage of illiterate adult women is found as 65% in Saudi Arabia. If we look at the Gender Parity Index (ratio of female to male) Qatar is again at the top with a GPI of 0.99 showing very close male—female parity in adult literacy.

The trend in youth literacy (i.e., literacy in the age group of 15–24 years) remains the same as in adults. Again, Kuwait is at the top with a 100% literacy rate in both the male and female categories, consequently the GPI is one. Another country having the same GPI level (i.e., unity) is Bahrain where the youth literacy rate is 97% in both the genders. Qatar has a GPI more than unity (1.03) showing a higher female literacy rate compared to male. The highest number of female illiterates is found in Saudi Arabia with 75% of females amongst the illiterate population.

Table 1. Adult and youth literacy rate.

	Adult (15 years and over)						Youth (15–24 years)					
	Literacy rate				Illiterate population		Literacy rate				Illiterate population	
GCC countries	Total	Male	Female	GPI	Total	(% Female)	Total	Male	Female	GPI	Total	(% Female)
Bahrain	87	89	84	0.94	66385	49	97	97	97	1	3359	43
Kuwait	93^{+1}	94^{+1}	91^{+1}	0.96^{+1}	138641^{+1}	49^{+1}	100^{+1}	100^{+1}	100^{+1}	1^{+1}	1094^{+1}	38^{+1}
Oman	81	87	74	0.85	300192	57	97	98	97	0.99	14356	59
Qatar	89	89	89	0.99	66686	29	96	95	98	1.03	4373	24
Saudi Arabia	79	87	69	0.8	2680976	65	96	98	94	0.96	157422	75
UAE	NA	NA	NA	NA	NA	NA	NA	NA	NA	NA	NA	NA

Source: Global Education Digest 2006: Comparing Educational Statistics across the World, UIS, June 2006.

Table 2. Numbers of universities in GCC countries.

GCC countries	2001/02	2002/03	2003/04	Current*
Bahrain	8	9	10	10
Kuwait	6	6	6	6
Oman	4	4	4	5
Qatar	3	4	5	5
Saudi Arabia	8	8	11	21
UAE	8	8	8	15
Total	37	39	44	62

Sources: 1. GCC Secretariat General Information Centre—Statistical Department, Statistical Bulletin Volume 15, 2006. 2. Information from Govternment websites of GCC Countries. (Details given in list of references.)

4 INFRASTRUCTURE SET-UP

The second dimension for the analysis of higher education is the Infrastructure setup in the GCC countries. Two important factors of this dimension will be studied, i.e. educational institutions and teaching staff. A good number of higher education institutions with quality teaching staff prepare the ground for quality in higher education. Let us first look at the university infrastructure in the GCC countries which is presented in Table 2.

The figures summarized in the Table 2, show an impressive growth in university infrastructure over a period of time with a total of 37 universities in the year 2001–2002 to 62 in 2007. The latest data on the number of universities were gathered by searching the relevant websites of GCC countries. The growth rate in the number of universities was 5.4% in 2002–2003, followed by 12.82% in the next year; the current growth rate is observed as 41%. If the number of universities is taken as a parameter for expansion of higher education infrastructure in these countries then Saudi Arabia takes a lead with most universities, 21, followed by the UAE with 15 universities. An interesting finding is that over the years, private and foreign universities in association with local institutions have played a vital role in the expansion of the network.

Apart from these regular universities there are a number of colleges and specialized institutions offering specific professional degrees and diplomas to the youth population of these countries.

The data pertaining to human resources availability in the universities are presented in Table 3. Looking at the availability of human resources, across the GCC countries, the availability of staff in the universities has been increasing constantly over the years along with the growth in the number of universities. While we have composite data on male and female teachers and their nationalities for three countries, namely Bahrain, Kuwait and Qatar, the data on male and female staff are only available for Saudi Arabia and the UAE, and data for Oman give only the total figure of university staff. No doubt, the countries having a higher number of universities also have a higher number of university staff catering to the educational needs of the society.

5 STUDENT ENROLMENT IN UNIVERSITIES

One of the most significant aspects evident from the higher education data is the continuous increase in the student enrolment in the universities over the years showing enhanced awareness amongst the youth towards taking up higher education and contributing towards development of the country. What is even more worthwhile noting here is the higher enrolment of the female segment of the population as compared to the male segment. This indicates greater participation of women in the higher education of the countries. As is evident in Table 4, Saudi Arabia and

Table 3. University staff by gender and nationality.

GCC countries	2001–02				2002–03				2003–04			
	Male		Female		Male		Female		Male		Female	
	Citizen	Non-citizen	Citizen	Non-citizen	Citizen	Non-citizen	Citizen	Non-citizen	Citizen	Non-citizen	Citizen	Non-citizen
Bahrain	241	154	217	84	245	173	267	117	240	164	367	137
Kuwait	492	394	166	21	529	429	248	62	560	386	193	17
Oman	1474				1490				946			
Qatar	153	294	148	62	149	295	142	59	154	310	151	61
Saudi Arabia	13403		8515		14213		9146		16794		8403	
UAE	190		2226		232		2716		NA		NA	

Source: GCC Secretariat General Information Centre—Statistical Department, Statistical Bulletin Volume 15, 2006.

Table 4. Student enrolment in university.

GCC countries	2001–02				2002–03				2003–04			
	Male		Female		Male		Female		Male		Female	
	Citizen	Non-citizen	Citizen	Non-citizen	Citizen	Non-citizen	Citizen	Non-citizen	Citizen	Non-citizen	Citizen	Non-citizen
Bahrain	5417		8770		8744		13347		10867		15273	
Kuwait	4632	579	10973	1027	4711	613	11435	1069	4848	616	11714	1125
Oman	9411	25	10455	42	9477	11435	10844	43	10001	45	10892	47
Qatar	1210	442	4786	899	1584	659	4925	977	1452	678	4702	1035
Saudi Arabia	196519		248281		219356		305988		236996		334817	
UAE	21751		42103		22550		44543		13939	10871	29767	15164

Source: GCC Secretariat General Information Centre—Statistical Department, Statistical Bulletin Volume 15, 2006.

the UAE again have the highest enrolment in universities of both males and females, followed by Oman.

Having focussed on the general enrolment trend in the universities, what follows here is specific attention to tertiary education levels in the GCC countries. This is based on the International Standard Classification of Education (ISCED 97). According to this system of classification of educational levels, the first stage of tertiary education comes under Level 5. This level is divided into two sub-categories, i.e., 5A and 5B. The ISCED 5A programmes are largely theoretically based and are intended to provide sufficient qualification for gaining entry into advanced research programmes and professions with high skills requirements. The following are the classification criteria for ISCED 5A programmes: minimum cumulative theoretical duration of three years; faculty having advanced research credentials; the programme involves a research project or thesis; and this level provides the level of education required for entry into a profession with high skills requirement or an advanced research programme.

The ISCED 5B programmes are generally more practical, technical, and occupationally specific than the 5A programmes. The classification criteria of these programmes are: they are more practically orientated and occupationally specific than the programmes of Level 5A and do not prepare students for direct access to advanced research programmes; they have a minimum two years' duration and programme content is typically designed to prepare students for a particular occupation.

The second stage of tertiary education is ISCED 6 level which leads to an advanced research qualification. These programmes are devoted to advanced study and original research. The programmes require submission of a thesis or dissertation which is a product of original research and contribution of specific knowledge. These programmes are not based solely on course work, rather it prepares participants for faculty posts in institutions offering five year level programmes, as well as research posts in government and industry.

The data in Table 5 are summarized on teaching staff, enrolment and graduates in tertiary education levels 5 and 6 in the countries of the GCC. As is evident from the Table 5, maximum enrolment in tertiary education is being observed in Saudi Arabia followed by the UAE, Kuwait, Oman, Bahrain and Qatar. If we compare the enrolment figure with the number of universities in GCC countries in 2003–2004, then we find that the student-university ratio is highest in Saudi Arabia (52157.45), followed by the UAE (8522.75), Oman (8451.75), Kuwait (7012.67), Bahrain (1852.40) and Qatar (1729.60), while the overall GCC ratio stands at 16931.11. Only Saudi Arabia has a higher student-university ratio than the overall GCC student-university ratio. Another aspect of the analysis is the student-teacher ratio. Qatar stands first if we look at the student-teacher ratio with a ratio of 11.73 (i.e. for every 12 students there is one teacher). The highest ratio is observed in Oman (29.55) while the overall GCC ratio is 23.03. Oman, the UAE (23.13) and Kuwait (25.61) have higher student-teacher ratios than the overall GCC ratio.

If we study the trend of graduating students and their field of graduation then we come across an interesting trend, but with some limitations because relevant data for the UAE and Kuwait are not available. In all the GCC countries more than half of the graduating students are female. The country and discipline analysis shows that Bahrain has the highest percentage (41%) of graduating students in Social Science, Law and Business. Oman has the highest ratio in the Education discipline (68%); Qatar has a similar as that of Bahrain (38%) in Social Science, Law and Business while in the case of Saudi Arabia again Education is the most sought-after discipline. If we look at the trend of female graduates in the GCC countries, we find that the performance of female students is far more impressive than their male counterparts. In Bahrain, in all the disciplines, female graduates have a share of at least 50% while in Oman it ranges from 37% to 60%. The analysis of Qatar shows a range of female graduates from 47% to 100% in different disciplines while in Saudi Arabia, the range is 29% to 78%. The school life expectancy in the GCC countries ranges from 0.6 (Oman) to 1.7 (Bahrain). The gender parity index has been studied for graduates in ISCED level 5 & 6 and it is very interesting to know that in all the GCC countries GPI is more than unity. This shows that female students are surpassing their male counterparts and showing a

Table 5. Enrolment, staff and graduates in tertiary education.

GCC countries	Enrolment and staff in tertiary education		Field wise graduates in tertiary education at ISCED level 5 & 6							
	Total enrolment in tertiary education in ISCED 5&6 in 2004 (% Female)	Teaching staff in tertiary education (% Female staff)	Total students (% Female)	S&T as % of total (% Female)	Education as % of total (% Female)	Humanities and arts as % of total (% Female)	Social Sc, business and law as % of total (% Female)	Health and welfare as % of total (% Female)	School Life expectancy in ISCED 5 and 6 level in years	GPI
Bahrain	18524 (63%)**	832 (36%)**	2555 (70%)$^{-1}$	19 (50)$^{-1}$	12 (79)$^{-1}$	7 (85)$^{-1}$	41 (69)$^{-1}$	12 (84)$^{-1}$	1.7 (2.3)$^{-1}$	1.84**
Kuwait	42076 (71%)**	1643 (23%)**	NA	NA	NA	NA	NA	NA	1.1 (1.7)**	2.72**
Oman	33807 (56%)	1144 (25%)	5059 (62%)	12 (38)	68 (72)	3 (41)	13 (37)	1 (60)	0.6 (0.7)**	1.37**
Qatar	8648 (67%)**	737 (31%)**	1386 (73%)$^{-1}$	16 (47)$^{-1}$	32 (90)$^{-1}$	10 (92)$^{-1}$	38 (62)$^{-1}$	4 (100)$^{-1}$	0.9 (1.5)**	3.05**
Saudi Arabia	573732 (59%)	25041 (34%)	81686 (53%)	9 (39)	41 (78)	28 (29)	15 (39)	6 (44)	1.4 (1.7)**	1.5**
UAE	68182 (66%)**$^{-1}$	2948**$^{-2}$	NA	NA	NA	NA	NA	NA	1.1 (2.0)**$^{-1}$	3.24**$^{-1}$

Source: Global Education Digest 2006: Comparing Educational Statistics across the World, UIS, June 2006.

Table 6. International mobility of students.

GCC countries	Total outbound mobile students (out bound mobility ratio)	Gross outbound mobility ratio	Total inbound mobile students	Net flow of mobile students
Bahrain	2108 (11.4)**	3.9**	1331**	–777
Kuwait	4959 (11.8)**	2.6**	NA	NA
Oman	4283 (12.7)	1.6	NA	NA
Qatar	1105 (12.8)**	2.3**	1633^{-1}	528
Saudi Arabia	9318 (1.6)	0.4	12199	2881
UAE	4384 (6.4)**	1.4**	NA	NA

Source: Global Education Digest 2006: Comparing Educational Statistics across the World, UIS, June 2006.

clear trend towards women emancipation and empowerment. The highest GPI is observed in Qatar and the least in Saudi Arabia.

6 INTERNATIONAL MOBILITY OF STUDENTS

Another dimension of higher education taken up for analysis is international mobility of the students in GCC countries. In Table 6, the international mobility of students is presented in terms of inbound mobile students (i.e., foreign students coming to the country for study) and outbound mobile students (native students going out of the country for study). However, we cannot have complete and detailed analysis and clear picture cannot emerge out of the analysis because data for the UAE, Oman and Kuwait are not available for inbound mobile students. Bahrain is the only country having more outbound mobile students compared to inbound mobile students, hence net flow in the case of Bahrain is negative. The highest positive net flow of students is observed in the case of Saudi Arabia with net flow of students at 2881. This shows that Saudi Arabia has the potential to attract students for higher education. A similar observation is found for Qatar also. This analysis is further strengthened with the study of outbound mobility ratio. The outbound mobility ratio is highest, 12.8, for Qatar and the least being 1.6 for Saudi Arabia. The UAE also has a good potential for retaining their native students for higher education in their country.

7 QUALITY OF EDUCATION STREAMS

Analysing another parameter of higher education, yet another picture emerges. The ranking of quality of education systems reveals that the position of Qatar is the highest amongst the GCC countries (i.e., 20) followed by the UAE which ranks in position 32 (see Table 7). The ranking is similar in the case of quality of mathematics and science education and that of business schools in these countries. However, there is a different portrait with respect to the availability of specialized research and training services in these nations. The UAE followed by Oman have better locally available specialized opportunities for research and training.

Having analysed and presented the data pertaining to the present-day higher education setup in GCC countries, an attempt is now being made here to make an internal as well as external analysis of the various factors influencing this setup. In this connection, a Strength Weaknesses Opportunity and Challenges (SWOC) analysis technique is utilized for determining the internal important factors affecting higher education. This will help in determining the emerging challenges faced by the higher education scenario in GCC countries.

Table 7. Quality of education systems.

GCC countries	Secondary enrolment rate (up to 2004)	Gross tertiary enrolment rate	Quality of education system education	Quality of mathematics and science schools	Quality of business	Local availability of specialized research and training services
Bahrain	98.8 (27)	34 (56)	3.2 (80)	3.4 (89)	3.7 (78)	3.1 (97)
Kuwait	89.9 (52)	22 (73)	3.5 (63)	4.1 (62)	4.2 (60)	4 (54)
Oman	86.4 (62)	13 (91)	4.2 (38)	4.3 (59)	4.2 (63)	4.3 (43)
Qatar	96.8 (34)	19 (77)	4.9 (20)	4.7 (38)	4.6 (40)	3.9 (58)
Saudi Arabia	NA	NA	NA	NA	NA	NA
UAE	66.4 (90)	22 (73)	4.4 (32)	4.5 (41)	4.4 (52)	4.3 (42)

Sources: 1. Global Education Digest 2006: Comparing Educational Statistics across the World, UIS June 2006; 2. Arab World Competitiveness Report 2007.

8 *SWOC* ANALYSIS

8.1 *Strengths*

These are

1. Existence of rich teaching resources in the Arabian world;
2. Favourable government policy towards establishment of institutions of higher learning in the private sector;
3. Existence of positive government framework for encouraging renowned foreign universities to set up their institutions in collaboration with local agencies or foundations; for instance, Texas A&M University and CHM University, Carnegie Mellon Campus in Qatar; American University of Sharjah and American University of Dubai, British University of Dubai in the UAE; AMA International University in Bahrain; American University of Kuwait, etc.;
4. Creation of a Special Education Zone (SEZ) in GCC countries promoting higher education especially of a professional nature in the country such as Knowledge Village in the UAE and Knowledge City in Qatar;
5. Existence of need-based and employment-centred programmes and institutions such as Medical University of Bahrain, King Fahd University of Petroleum and Minerals, King Saud bin Abdul Aziz University for Health Sciences in Saudi Arabia, Gulf University for Science and Technology in Kuwait, Ajman University of Science and Technology in the UAE, etc.;
6. Internationally-accredited educational programmes in the field of health and medical science, engineering and technology, computer science and information technology, business studies and law, public administration, pure science, and allied social sciences such as anthropology, psychology, etc.;
7. Attractive salary and service benefits for teaching and research professionals in the region.

8.2 *Weaknesses*

Weaknesses include

1. Reliance on non-citizen expertise for teaching in specialized and professional courses;
2. Limited professional courses for enhancing job opportunity for women;

3. Need for long term planning for integrative and networked educational development;
4. Perceived gap between need of the job market and educational output.

8.3 *Opportunities*

The opportunities presented include

1. Need for development of an integrated and networked university in GCC countries that would offer region-based, employment-centred academic and professional programmes for GCC nationals exclusively;
2. Exchange of faculty, expertise and teaching resources and students amongst the universities of GCC countries;
3. Development of comprehensive training programmes for teachers and researchers uniformly in all the GCC countries;
4. Greater and wider implementation of ICT in making educational resources available to students across the region;
5. Considering the large number of students at the primary and secondary level institutions, there is a need to establish more quality higher educational institutions so as to give greater opportunities to both male and female students;
6. In view of the rising awareness amongst women towards attaining higher education and also greater support provided to them by society by providing them with education and empowering them, educational institutions with job-oriented programmes specifically for women need to be set up;
7. Increasing demand for expansion in higher educational institutions;
8. Establishment of educational institutions in public-private partnership through international collaboration.

8.4 *Challenges*

The challenges facing educators in GCC countries include

1. Self-sufficiency in developing qualitative human leadership
    a. Professionalization in not only teaching personnel but also in educational leaders such as supervisors, principals, educational advisors, laboratory and workshop technicians, librarians and their counterparts in the university education setup;
    b. Adoption of a licensing system for the teaching profession;
    c. Need for comprehensive and integrative teacher training and development programme;
    d. Training and development for senior educational administrators including university faculties for strategic thinking, planning and administration.
2. Organizational challenges
    a. Linking of teaching function with scientific research and development;
    b. Development of procedural system for supervision, evaluation and accountability and comprehensive assessment system of educational performance.
3. Financial challenges
    a. Finding other financial alternatives that can assist the government in the expansion of educational institution and fulfilling the requirement of education system development.
4. Efficiency and effectiveness of educational outputs
    a. Developing science, mathematics, and other employment-centred academic programmes such as IT, bio-technology, bio-chemistry, agriculture science, business studies, etc., so that employability of students could be improved;

b. Devising attractive academic programmes leading at tertiary level leading to enhanced employability for non-mathematics and non-science-background pass-outs at secondary or senior secondary level;
c. Bridging the gap between quality of higher education output and the needs of the job market;
d. International quality standards in the education system so that the out-bound mobility rate of students could be minimized. Simultaneously make the educational set up much more attractive for enhancing the in-bound mobility of the students to the GCC countries;
e. Design and development of modular programmes in English language which can cater to the needs of the students belonging from non English-speaking backgrounds. This should be able to bring these students to a certain level of language competency and general awareness.

9 THE WAY AHEAD: GCC MAPPING OF HIGHER EDUCATION

Having considered and analysed the present educational scenario of GCC countries and also highlighted the strengths, weaknesses, opportunities and challenges, an effort is being made here now to develop a framework of GCC Mapping of Higher Education and suggest strategies for facing the challenges, thus fulfilling the objectives of the Principle Theme of the Conference, i.e., "Bahrain as a Hub for Higher Education"

This policy document "GCC Mapping of Higher Education" has been prepared on the following assumptions:

a. There is a need for a collaborative and integrated **GCC University** having a mandate to provide higher education of international standard at par in all the GCC countries;
b. The University may have headquarters in one of the GCC countries and Regional headquarters in all the member countries of the GCC;
c. The level and nature of academic curriculum and delivery of education programmes should be the same in all the GCC countries and complete decentralization of operation of the university under proper guidance of GCC University Headquarters;
d. Centre of Excellence in each of the member countries to undertake research and development activities related to respective specialization areas and design and development of academic programmes.

9.1 *Structure of the GCC university*

A. Keeping to the theme of the Conference, the authors hereby propose that the GCC University Headquarters should be in Bahrain and the Regional Headquarters in Manama, Riyadh, Doha, Kuwait, Muscat and Dubai;
B. Centre of Excellence (COE) to be created in all the member countries depending upon their strength in respective specialization areas as *per* the following details;

   a. Centre of Excellence in Petroleum and Minerals: Riyadh, KSA;
   b. Centres of Excellence in Tourism , Hospitality, Agriculture and Construction: Dubai, Sharjah, Abu Dhabi and Ajman in the UAE;
   c. Centres of Excellence in Natural Gas, Business Management and Science and Technology: Doha, Qatar;
   d. Centres of Excellence in Chemical Engineering, Textiles, Health and Family Welfare: Muscat, Oman;
   e. Centres of Excellence in Banking and Finance, IT & Tele-Communication, Housing and Recreation : Manama, Bahrain;
   f. Centres of Excellence in Food Processing, Bio-Technology, Bio-Chemistry and Environmental Science: Kuwait.

C. Key Responsibility Areas for Centres of Excellence:

a. Identification of need and employment potential of educational programmes with respect to Centres of Excellence;
b. Design, development and delivery of these academic programmes under proper co-ordination and collaboration with GCC University headquarters and regional headquarters;
c. Training, development and capacity enhancement of teaching and research professionals in these specialization areas;
d. Drawing professionals from the member countries of the GCC and their training and development as *per* the University requirement;
e. Performance evaluation of personnel;
f. Identification of potential for developing ICT based online and networked learning academic programmes for GCC nations.

D. Expected Achievements of GCC University:

a. Reducing outbound mobility of students and increasing inbound mobility of students of member countries;
b. Reducing reliance of students on foreign universities for quality higher education;
c. Potential for drawing experts across the globe and gradually developing an in-house pool of experts;
d. Gradual self-reliance in teaching and research over a period of time;
e. Education and curriculum at par with international standards;
f. Tailor-made academic programmes based on the needs of GCC countries;
g. Retention of educational talent in GCC countries;
h. Centralized placement division with regional placement cells in all the member countries that would provide necessary support for placement of student pass-outs.

E. Proposed Action Plan for GCC Countries:

a. Decentralized and uniform admission standards in the GCC University.
b. Decentralization of human resource planning, management and development at the Regional headquarters level;
c. Imparting specialized training to teaching professionals of regional headquarters in Centres of Excellence;
d. Centralized career planning and progression of University staff;
e. Temporary secondment of expertise from Centres of Excellence to different regional headquarters;
f. Bi-annual congregation of experts at the respective Centres of Excellence to review progress and policy formulation;
g. Annual performance evaluation of staff at the University headquarters.

F. Financial Resources: The plan for the GCC University is to have public-private partnerships for mobilization of funds from the public and private sectors. The public sector organizations would include GCC Secretariat, respective Governments of GCC countries, UNESCO, and the public sector majors of GCC countries while fund mobilization from the private sector would include corporate bodies and leading industrial organizations in order to fulfil their social responsibility.

As a part of the GCC mandate and commitment of the member countries towards universalization and provision of quality higher education, it is the need of this decade to provide an adequate platform to the youth for attaining and upgrading their levels of knowledge and competency. This should be available to all concerned, belonging to all walks of life in order to bring a developmental parity amongst all the member nations. The time has come for all of us to understand and realize that a knowledge society is developing in the 21st century.

GLOSSARY

Key symbols used in tables are as follows:
* national estimation,
** UIS estimation,
– magnitude nil or negligible,
. not applicable;
+*n*: data referred to the school or financial year (or period) *n* years or periods after the reference year or period;

- *n*: data referred to the school or financial year (or period) *n* years or periods before the reference year or period;
- All the ratios are expressed as percentages except the GPI (Gender Parity Indices);

UIS: UNESCO Institute of Statistics.

REFERENCES

Reports:

1. Global Education Digest, 2006. *Comparing Educational Statistics across the World*, UIS June 2006.
2. GCC Secretariat General Information Centre—Statistical Department, Statistical Bulletin Volume 15, 2006.
3. World Economic Forum, 2007. *Arab World Competitiveness Report 2007.*
4. GCC Secretariat, Comprehensive Development of Education in GCC Countries: A Study of the Directives Stated in the Supreme Council Resolution, 23rd Session, (Doha, December 2002) on Education.
5. UNESCO Institute of Statistics, *Annual Report 2003–2004*, Montreal, Canada.

Websites:

1. www.the-saudi.net/directory/sag/htm
2. www.bahrain.gov.bh
3. www.e.gov.qa
4. www.oman.org/gov00.htm
5. www.government.ae/gov/en/biz/index/jsp
6. www.gksoft.com/govt/en/kw.html
7. http://www.gcc-sg.org/

Higher Education in the Twenty-First Century: Issues and Challenges – Al-Hawaj, Elali & Twizell (eds)
© 2008 Taylor & Francis Group, London, ISBN 978-0-415-48000-0

Delivering quality higher education through e-learning: A conceptual view

A. Usoro & A. Abid
University of the West of Scotland, Paisley, U.K

ABSTRACT: Both academic and non-academic institutions such as businesses have increasingly been interested in the use of information and communication technologies (ICT) to support learning otherwise termed e-learning. This interest has been fuelled by the new developments in ICT such as multi-media and the Internet with its worldwide web. Other incentives have been the associated (expected) reduction of the cost of education and the easier expansion of education to the increasing market that cannot be reached by traditional delivery. Especially with higher education (HE), the issue of quality is raised leading to both anecdotal and empirical evidence of ways to maintain quality while deriving the benefits of e-learning. This paper aims to discuss the issue of quality in higher education and examine how it can be maintained in online learning. It will present key current research as well as highlight possible areas for future studies.

1 INTRODUCTION

It is not yet possible to be conclusive on what constitutes quality in e-learning because it is only in the 1990s that the expansion in information and communications technologies, especially the Internet, has motivated the explosion of research and practice in this field. Indeed, growth in technology has run at a faster pace than research and full understanding of what should be quality in e-learning. Thus, the International Standards Organisation (ISO) is yet to finalize framework standards for quality e-learning. Given this background, this paper attempts to explore the concept of quality in e-learning by reviewing literature in higher education and e-learning. Based on this desk research, an attempt is made to identify common themes and to summarize what may be regarded as quality criteria in e-learning and pointers are made of areas for further investigation.

2 OPERATIONAL DEFINITIONS

2.1 *Quality in higher education*

The change factors described by Green (1994) below have fuelled the current interest in quality of HE in developed countries:

- rapid expansion of student numbers in the face of public expenditure worries;
- the general mission for better public services;
- increasing competition within the educational "market" for resources and students; and
- tension between efficiency and quality.

Interest in quality higher education is expressed not only in developed economies, like the UK, but also in less developed ones. For instance (Idrus, 1999) discusses efforts towards achieving quality higher education in Indonesia.

Granted that educational institutions are not pure business organizations that are swiftly changing with the environment, there has been some paradigm shift in the concept of higher education (Pond, 2002) as indicated in Table 1.

Table 1. Old and new concepts for accreditation and quality assurance.

Old Paradigm	New paradigm
Teacher/institution centred	Learner centred
Centralized	Localized
Homogenous	Deferential
One size fit all	Tailored
Closed	Open
Us *versus* them	Collaborative
Prescriptive	Flexible
Time as constant/learning as variable	Learning as constant/time as variable
Teacher credentials	Teacher skills
Consolidated experience	Aggregated experience
Regional/national	International/global
Static	Dynamic
Single delivery model	Distributed delivery model
Process	Outcomes
Infrastructure	Services

Source: Adapted from Pond (2002).

As Ellis et al. (2007) have stated, quality for higher education is a vexed term as it may suggest the notion of accountability at the expense of improvements. Yet, quality has to be conceptualized in order to improve it. Higher education belongs to the service rather than the manufacturing industry which has more precise measures for quality. This has not discouraged attempts at conceptualizing and even measuring this concept in higher education (cf Ashworth and Harvey, 1994; Harvey et al., 1992). Owlia and Aspinwall (1996) studied earlier attempts and also quality frameworks in other disciplines such as software engineering which they argued are akin to higher education. From their study they produced a conceptual framework that groups 30 attributes into the six dimensions of tangibles, competence, attitude, content, delivery and reliability (see Table 2).

Owlia and Aspinwall's (1996) research appears to be the most comprehensive dimensioning study of quality of higher education. Other studies appear to confirm, complement all or some quality dimensions of Owlia and Aspinwall's research. Their study also recognizes that quality is in the eye of the beholder. They argued that interests of stakeholders vary with the quality dimensions. For instance, while content and reliability are of interest to students, staff and employers, tangibles and competence are not of interest to employers; and attitude and delivery are of interest to students only.

The rest of this section will present a few additional studies and quality standards which can be seen to support Owlia and Aspinwall's dimensions.

The International Organization for Standardization (ISO) (2007) has defined its HE quality criteria as:

- Content and pedagogical method;
- Achievements and impact of the programme demonstrated by performance indicators (number of students, number of nationalities, assessment and follow up of students, profile of teachers, etc.);
- Connection of the programme to business, governmental and other stakeholder groups;
- Replicability of the programme—whether it could be implemented elsewhere in the world;
- Visibility of the programme, in particular in the media.

Table 2. Quality dimensions in higher education by Owlia and Aspinwall (1996).

No	Dimensions	
1	Tangibles	Sufficient equipment/facilities Modern equipment/facilities Ease of access Visually appealing environment Support services (accommodation, sports, …)
2	Competence	Sufficient (academic) staff Theoretical knowledge, qualifications Practical knowledge Up to date Teaching expertise, communication
3	Attitude	Understanding students' needs Willingness to help Availability for guidance and advice Giving personal attention Emotion, courtesy
4	Content	Relevance of curriculum to the future jobs of students Effectiveness Containing primary knowledge/skills Completeness, use of computer Communication skills and teamworking Flexibility of knowledge, being cross-disciplinary
5	Delivery	Effective presentation Sequencing, timeliness Consistency Fairness of examinations Feedback from students Encouraging students
6	Reliability	Trustworthiness

The recognition of the increasing globalization of higher education is reflected in some of the ISO items such as number of nationalities of students and the ability to implement a course in more than one place in the world. Like Owlia and Aspinwall, ISO standards accept that there are a number of stakeholders whose views have to be countenanced to have a holistic measure of quality. However, investigating quality education from the students' perspective is increasingly becoming popular by both higher education institutions and external stakeholders including the quality authorities (HEFCE, 2002; QAA, 2002). We will now look at two such studies.

Hill and Lomas (2003) are two researchers who examine quality from the students' perspective. They found the most influential quality factors in the provision of higher education to be the quality of the lecturer and the student support systems (Hill and Lomas, 2003). Their research was carried out on nursing, management studies, and learning and teaching students. The main question they asked was "What does quality education mean to you?" Other factors that emerged from Hill and Lomas' (2003) research are social/emotional support systems, and resources of library and IT.

Concentrating on the two most prominent factors, quality of the lecturer requires the lecturer to know his/her subject, be well organized and interesting to listen to by being positive and enthusiastic. These requirements pertain to the delivery in the classroom. The relationship with students in the classroom adds to the quality of the lecturer. The relationship appreciated by the students has to be easy and helpful for learning. Anderson (2000) agrees with the importance of the quality of the lecturer. Large class sizes and the increasing use of e-learning and resource-based learning are feared to affect the required stimulating and enthusiastic environment between the lecturer and the students (Gibbs, 2001). These fears necessitate quality assurance in e-learning environments.

Lagrosen and Seyyed-Hashemi (2004) are two researchers who investigated quality from the students' perspective. After reviewing earlier studies of quality in higher education, they used interviews and questionnaires to study 448 Austrian and Swedish students and their study revealed as significant the following factors:

- Courses offered;
- Computer facilities;
- Information and responsiveness;
- Collaboration and comparison.

Students' perspective is of course an important piece of the jigsaw but not the whole picture. Otherwise, we would miss quality research which should underpin HE. Thus, it is necessary to have academic and other stakeholders' views to achieve a comprehensive measure of quality. Moreover, a factor that is most often ignored given the regulatory and competitive demands is the quality of the students. Though achievement of entry qualifications, e.g., The Graduate Management Admission Test® (GMAT), have been proven to positively correlate with success in completion of education programmes, the increasing commercialization of higher education and increasing competition have made a few institutions to lower or entirely remove the needed entry qualifications with the results that weaker "inputs"—students—are used with the aim of providing quality "outputs"—graduates. This has the great potential of weakening quality education (Wilson, 2007). The existence of such a weakening effect places great responsibility on increasing the quality (and perhaps quantity) of the student support system.

Though it is a few years since the study was carried out, parts of Owlia and Aspinwall's (1996) framework maps to subsequent studies and views (or the reverse) on quality of higher education as shown in Table 3.

We will see that the quality dimensions of HE will appear in difference guises in various e-learning research that will be presented below.

2.2 *E-learning*

Since the late 1990s, there has been tremendous interest on e-learning both by practitioners and academics. E-learning is used to deliver training, education and collaboration using various electronic media but predominantly the Internet the tools of which have constituted the main driver of e-learning (Kramer, 2000). If it is well designed and managed e-learning can overcome many barriers associated with traditional learning which include students' tardiness, schedule conflicts, unavailable courses, geographical isolation, demographic and economic disadvantage (Hijazi, 2004).

2.3 *What is E-learning?*

The "e" in e-learning stands for "electronic" and just like similar terms, for instance e-business stands for computer mediated activities, e-learning refers to learning with the use of communication and information technologies. This agrees with the definition by the Higher Education Funding Council of England (HEFCE, 2005): "any learning that uses ICT" (p. 5). Although this definition is broad enough to cover non-online, e.g. CD, media, the internet has so dominated the actual implementation of e-learning that authors such as Gunasekaran et al. (2002) simply define e-learning as "Internet-enabled learning". If we take the broad sense e-learning, which

Table 3. Approximate mapping of various HE quality frameworks.

No	Owlia and Aspinwall (1996)	International Organization for Standardization (2007)	Hill and Lomas (2003)	Lagrosen and Seyyed-Hashemi (2004)
1	Tangibles		Resources of library and IT	Computer facilities
2	Competence	Profile of teachers	Quality of lecturer	
3	Attitude	Follow-up of students	Student support. Social and emotional support systems	Information and responsiveness. Collaboration and comparison
4	Content	Content and pedagogical method		Courses offered
5	Delivery	Assessment and follow-up of students		Collaboration and comparison
6	Reliability			
7		Globalization in terms of number of nationalities of students and replicability of programme		

is the use of ICT to support students in achieving their learning outcomes, a mixed strategy that combines the traditional approach to learning is accommodated (Ellis et al., 2007). Thus, we can observe many universities with face-to-face contact with students, using e-learning tools like Blackboard to make learning materials available to students and to co-ordinate their learning activities. In the UK, most of the successful e-learning programmes are the blended rather than the pure (no face-to-face contact) approach. Ennew and Fernandez-Young (2006) report on the failure of the flagship UK e-University (UKeU) project pointing out the need not to underestimate demand analysis in any major e-learning venture.

The argument as to whether e-learning is superior to face-to-face learning is not settled but e-learning has entered the centre stage in today's delivery of training and education whether in industry or in educational institutions. Players in the e-learning field are varied and include content providers, technology vendors and service providers, and they target academic, corporate and consumer markets. Department of Labour projected an increase from USD 550 million (BHD 206.8 million) to USD 11.4 billion (BHD 4.29 billion) in corporate e-learning revenues; and this increase represents 83% compound annual growth rate (Gunasekaran et al., 2002).

2.4 *Pros and cons of e-learning*

A number of reasons make e-learning appealing to educational and non-educational institutions as well as learners. The wide acceptance and availability of the internet means e-learning eliminates learning barriers of time, distance, socio-economic status while at the same time allowing individuals to take more responsibility for their learning which can now become life-long. Students can have access and benefit from a variety of experts and resources that may not be available locally. The fact that materials and delivery can easily and very economically be replicated is a strong appeal to higher education which seeks lower costs in the face of increased learning demands (Alexander, 2001; Antonucci and Cronin, 2001, Gunasekaran et al., 2002, Hijazi, 2004; Osborne and Oberski, 2004). E-learning mode of educational and training delivery along with its collaborative tools can transform a non-learning organization into a learning one, which is a very desirable attribute especially in the current global environment. A learning organization has the additional advantage of boosting staff morale and motivation (Tarr, 1998).

There is also a change in the way books are now published in the internet age. Authors provide plenty of electronic materials which include extra examples, problems and solutions, animated simulations, user group and feedback system, bank of test and examination questions with solutions, tools to compose exams and texts, and provision for lecturers to design their courses and customize material to their students. Lecturers also appreciate other tools that allow them to manage their students and their learning experience. The fact that these are all available on the internet also means that lecturers have easier access to a variety of publications with the associated support materials as explained in this paragraph.

Research evidence exists of the advantages of e-learning in terms of giving teachers access to more resources, being an effective way to implement national curriculum and instruction standards, and producing better performed learners than the traditional approach (Larson and Bruning, 1996; McCollum, 1997)

This rosy picture does not, at least at the moment, eliminate challenges of motivating lecturers and students to take up e-learning. Nor does it remove the challenges of achieving high quality learning by, for instance, providing a high stimulation learning experience. Indeed, Alexander and McKenzie (1998) reported the following as reasons for e-learning failure:

- Overly ambitious desired outcomes given the budget and time constraints;
- Use of technologies with disregard of appropriate learning design;
- Failure to change learning assessment to match changed learning outcomes;
- Software development without adequate planning; and
- Failure to obtain copyright clearance.

Other problems with e-learning are:

- Initial costs may exceed costs of traditional methods
- More responsibility is placed on the learner who has to be self-disciplined especially in the pure e-learning mode where there is no face-to-face interaction
- Some learners have no access to computer and/or the internet or the technology they use may be inadequate
- Increased workload (Connolly and Stansfield, 2007)
- Non-involvement in virtual communities may lead to loneliness, low self-esteem, isolation, and low motivation to learn, and consequently low achievements and dropout (Rovai, 2002)
- Dropout rates tend to be 10 to 20 percent higher than traditional face-to-face programmes (Carr, 2000).

These problems bring the issue of quality in e-learning to the fore.

2.5 *Quality in E-learning*

We cannot begin to think about the quality of e-learning if we cannot firstly convince a significant number of academic staff to take up the new technology in the first place. Despite the potentials of e-learning, a number of studies have identified lecturers' attitude as a barrier to the acceptance of e-learning (Pajo and Wallace, 2001; Sellani and Harrington, 2002; Newton, 2003). From an analysis of earlier work in this area, followed by empirical study, Newton (2003) identified the following barriers associated with lecturers:

1. Increased time commitment (workload)
   - o Development time,
   - o Delivery time;
2. Lack of incentives or rewards;
3. Lack of strategic planning and vision;

4. Lack of support
   - o Training in technological developments,
   - o Support for pedagogical aspects of developments;
5. Philosophical, epistemological and social objections.

These issues have to be sorted out to allow the free flow of quality e-learning. Besides lecturers, some potential students and employers are cautious of, and many national governments refuse to give full recognition to e-learning qualifications (Ennew and Fernandez-Young, 2006, p. 150).

At the same time, effective quality strategies, initiatives and tools are very important in convincing lecturers and other stakeholders to adopt e-learning (Gunasekaran and McNeil, 2002). Friesner (2004) has reported on the threat of plagiarism and poor academic practice to e-learning extension in higher education irrespective of country. The need to overcome this threat has to be recognized in a sound framework for quality e-learning in HE. Such confidence would go a long way towards building, in the stakeholders, confidence in e-learning.

Much research interest has developed and there are a few initiatives towards developing a sound framework for e-learning quality. The proposals of existing research and initiatives can be approximately mapped into the HE quality dimensions already discussed (for the mapping see Table 2). The rest of this section will present some existing research and initiatives.

The International Organization for Standardization (ISO) from early 2006 has been developing a framework for standards to reduce the cost and complexity of adopting quality e-learning approaches and, at the same time, to facilitate the introduction of new e-learning products and services (Training Press Releases 2006). The framework will introduce:

- Quality model to harmonize aspects of quality systems and their relationships;
- Reference methods and metrics to harmonize methods used to manage and ensure quality in different contexts,
- Best practice and implementation guide to harmonize criteria for identification of best practice, guidelines for adaptation, implementation and usage of multi-pronged standard. The guide will also contain best practice examples.

Targeting individualized adaptability and accessibility in e-learning, education and training, the International Organization for Standardization (ISO, 2007) is also currently[1] developing the following 8-part e-learning framework:

Part 1: Framework and reference model;
Part 2: "Access for all" personal needs and preferences;
Part 3: "Access for all" digital resource description;
Part 4: "Access for all" non-digital resource description;
Part 5: Personal needs and preferences for non-digital resources;
Part 6: Personal needs and preferences for description of events and places;
Part 7: Description of events and places;
Part 8: Language accessibility and human interface equivalencies (HIEs) in e-learning applications.

The benchmarks provided by the US Institute (Gunasekaran et al., 2002, p. 48) for ensuring e-learning quality and evaluating higher education effectiveness and policy include:

- A documented technology plan that includes password protection, encryption, back-up systems and reliable delivery;
- Established standards for course development, design and delivery;
- Good facilitation of interaction and feedback; and
- The application of specific standards for evaluation.

[1]As at May 2007.

The key issues so far seem to be the technology including easy access, course content as evidenced in good course development and design, delivery, and good support system. We will observe these issues repeated in the rest of the presented literature.

While acknowledging the existence of useful tools, like Blackboard, for e-learning, Webb (2000) considers the following non-technical issues in developing his on-line learning course:

1. Will the course be able to stand alone as a valid learning experience for the different student profiles that will be exposed to it? It may be necessary to supplement with non electronic activities.
2. Are the objectives, goals and assessment methods clear and compatible with e-learning format?
3. What remedial action will you take to care for non-performers in a timely fashion?
4. How will mentoring and guidance be carried out?
5. Beyond course completion, what will be the success criteria?

The framework proposed by Boticario and Gaudioso (2000) is:

1. Developing an interactive and on-line resource model which considers lecturers, students, tutors and other stakeholders at various levels;
2. Developing significant and active learning by stimulating student participation in the various learning resources;
3. Providing individualized communication to learners who are also given individualized and quick access;
4. Develop a "community of practice" with capability for knowledge sharing and collaboration among learners. Such development may go beyond the formal learning to provision of informal facilities like a "chat room".

To provide adequate support for students the following issues should be addressed (Newton, 2003, p 420):

- Advanced course requirements information;
- Close personal interaction;
- Equivalent library materials and research opportunities;
- Equally rigorous assessment as in campus-based learning;
- Academic counselling and advice;
- Handling of plagiarism, authentication and online academic misconduct.

In his contribution, Thomas (1997) proposes the following key elements:

1. Provision of learning materials;
2. Provision of facilities for practical work (e.g., simulation);
3. Enabling questions and discussions;
4. Assessment;
5. Provision of student support services (advising)

While, there may exist general frameworks for quality e-learning, it has to be acknowledged that courses differ and in their interactive and simulation requirements. The differences are well recognized in Gunasekaran et al.'s (2002) proposal of e-learning application shown in Table 4.

Alexander (2001) in his contribution complements existing work by tackling strategic issues and putting policies in place to faculty members involved in e-learning. His proposals can be summarized as follows:

- Development of a vision of e-learning;
- Development of a technology development plan;
- Development of faculty workload policies which takes into consideration the demands of e-learning;
- Maintenance of a reliable technology network;

Table 4. Proposal of e-learning application.

Learning Areas	Internet applications	E-learning strategies and technologies
Arts	Online classes for arts classes such as language, improving vocabulary and writing skills	E-mail, interactive and animated video on the WWW, exchanging files
Business	Business courses on the Internet, group projects, virtual company tours	E-mail, WWW, chat room, news groups
Engineering	Engineering classes on-line, virtual laboratory, virtual design, team learning and group projects	E-mail, WWW integrated CAD, hyperlinks, and 3D navigation
Science	Virtual laboratory, design of experiments, collaborative projects	E-mail, WWW, Internet chat room
Medicine	Simulation of surgical operations, diagnosis, cat room	WWW, WebMD, Internet
Agriculture	Treatment of crops from time to time, training and education using WWW	E-mails, WWW for training and education, multimedia application
Law and justice	Practice of law online, communication, simulation games	EDI, EFT, WWW and Internet

- Provision of facility for technical support for both staff and students;
- Market research support; and
- Provision of time release for faculty members engaged in e-learning developments.

Zhao (2003) summarized existing quality of e-learning research into his proposed framework which has the following components:

- Course effectiveness;
- Adequacy of access in terms of technological infrastructure;
- Student satisfaction.

The components of quality in e-learning that have been presented overlap each other. A proposed amalgamation of the key research and initiatives into quality of e-learning is shown in Appendix A.

The student experience with e-learning may be different from the traditional face to face delivery (Ennew and Fernandez-Young, 2006, p. 150) which causes fears of less student stimulation with e-learning. However, there is no relenting in effort to animate and introduce interactivity in e-learning: scripting languages like Java and programming paradigms like second life (Cross, 2007) have been used. Also, it has been observed that the skills developed when playing computer games are useful for e-learning and in some cases the same skills are the very objectives (e.g., quick development of strategies in reactions to one's environment) of learning (Gee, 2004). Thus, some research effort is directed at the application of computer games in e-learning. An example is Connolly and Stansfield (2007).

3 AREAS FOR FURTHER RESEARCH

The research initiated by this paper is obviously work-in-progress. The mapping of the quality dimensions need to be validated and refined by primary research so that an empirically tested framework can be presented for the introduction and management of e-learning in higher education.

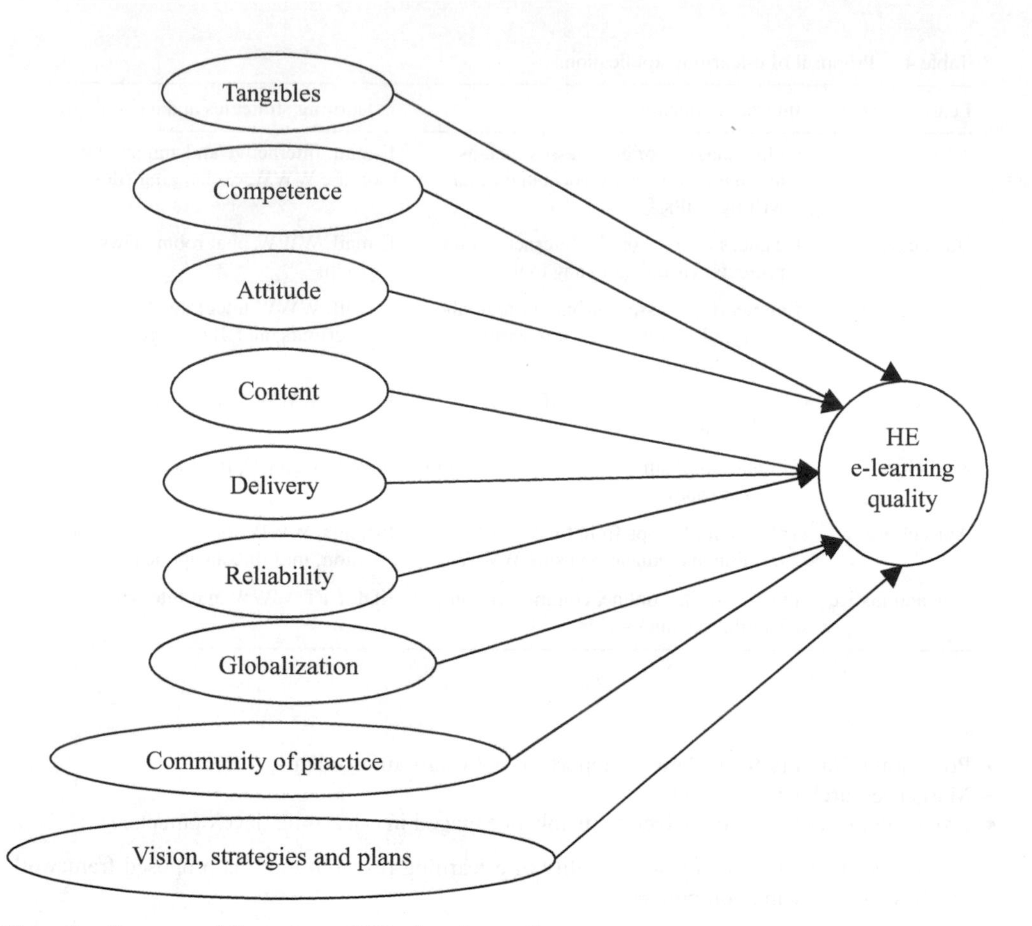

Figure 1. Summary of dimensions of HE e-learning quality.

There is also another interesting area for further investigation: while e-learning promises to save time (because of the technological ease of replication, access to learning materials and communication) increased workload has been identified as the negative effect of taking up e-learning. Does e-learning save time for the lecturer? Are there intervening factors, like phase in development, e-learning skills and experience of the lecturer that decides whether e-learning saves or consumes time?

Furthermore, considering Figure 1, it would be interesting to find out whether there are any relationships between the proposed dimensions of e-learning on the one hand and the HE e-learning quality on the other (and, if they exist, what the significance of such relationships are).

4 SUMMARY AND CONCLUSION

There are overlaps in existing research and initiatives towards conceptualizing and dimensioning quality in higher education as well as in e-learning. Drawing from secondary study and using HE quality framework by Owlia and Aspinwall (1996) as the base, this paper attempts to summarize the HE quality components into tangibles, competence, content, delivery, reliability and globalization. The components have been used as a basis to map research into dimensions of e-learning quality (see Appendix A). Extra components are (1) creation of communities of practice, and (2) development of e-learning vision, strategies and plans. The dimensions have been summarized in the Figure 1.

It has to be borne in mind that the varied perspectives (from students, academics, employers and other stakeholders) on the quality dimensions have to be countenanced to arrive at comprehensive measure of quality.

One of the major concerns of quality is the challenge of making computer-assisted learning to be as stimulating as the traditional face-to-face mode of learning. This concern has motivated the use of simulated computer environments such as those provided by computer games, and research is still in progress on how best to utilize these facilities and improve the e-learning experience. There are also key strategic and policy issues of motivating staff and giving them both the technical and pedagogical support to undertake e-learning.

It is proposed that paying adequate attention to these factors would help higher education institutions to achieve high quality in e-learning. Nonetheless, future research will empirically test these factors.

REFERENCES

Alexander, S & McKenzie, J. 1998. An evaluation of information technology projects in university learning. Department of Employment, Education and Training and Youth Affairs. Australian Government Publishing Services. Canberra.

Antonucci, R.V. & Cronin, J.M. 2001. Creating an on-line university. *Journal of Academic Librarianship*, 27(1): 20–23.

Ashworth, A. & Harvey, R.C. 1994. *Assessing Quality in Further and Higher Education*. London: Jessica Kingsley.

Alexander, S. 2001. E-learning developments and experiences. *Education and Training*, 43(4,5): 240–248.

Anderson, L. 2000. Teaching development in higher education as scholarly practice: a reply to Rowland et al. turning academics into teachers. *Teaching in Higher Education*, 5(1): 23–31.

Boticario, J.G. & Gaudioso, E. 2000. Adaptive web-site for distance learning. *Campus-Wide Information Systems*, 17(4): 120–128.

Carr, S. 2000. As distance education comes of age, the challenge in keeping students. *Chronicle of Higher Education*, 46(13): A39–A41.

Connolly, T.M. & Stansfield, M. 2007. From e-learning to games-based e-learning: using interactive technologies in teaching on IS course. *International Journal of Information Technology and Management*, 6(2,3,4): 188–207.

Cross, J. O'Driscoll, T. & Trondsen, E. 2007. Another life: virtual worlds as tools for learning. *e-Learn Magazine*, 2007(3): 2.

Ellis, R.A., Jarkey, N. Mahony, M.J., Peat, M. & Sheely, S. 2007. *Quality Assurance in Education*, 15(1): 9–23.

Ennew, C.T. & Fernandez-Young, A. 2006. Weapons of mass instruction? The rhetoric and reality of on-line learning. *Marketing Intelligence and Planning*, 24(2): 148–157.

Gee, J.P. 2004. *What Video Games Have To Teach Us About Learning and Literacy*. Hampshire: McMillan.

Gibbs, P. 2001. Higher education as a market: a problem or solution. *Studies in Higher Education*, 1: 85–94.

Green, D. 1994. What is quality in higher education? in: *Concepts, Policy and Practice*: 3–20, Open University Press, Milton Keynes, U.K.

Gunasekaran, A., McNeil, C.R. & Shaul, D. 2002. E-learning: research and applications. *Industrial and Commercial Training*, 34(2): 44–53.

Hart, M. 2004. Plagiarism and poor academic practice—a threat to the extension of e-learning in higher education? *Electronic Journal of e-Learning,* 2(2): Paper 25.

Harvey, L., Burrows, A. & Green, D. 1992, in: *Criteria of Quality: Summary*, The University of Central England, Birmingham.

Higher Education Funding Council for England (HEFCE) (2002). Information on Quality and Standards in Higher Education (02/15). HEFCE. Bristol.

Higher Education Funding Council for England (HEFCE) (2005). HEFCE Strategy for e-Learning. JISC, HEA. London.

Hijazi, S. 2004. Interactive technology impact on quality distance education. *Electronic Journal of e-Learning*, Vol 2, Issue 2, December, paper 5.

Hill, Y. & Lomas, L. 2003. Students' perceptions of quality in higher education. *Quality Assurance in Education*, 11(1): 15–20.

Indrus, N. 1999. Towards quality in higher education in Indonesia. *Quality Assurance in Education*, 3(3): 134–141.
ISO (2007) ISO/IEC FDIS 24751 http://www.iso.org/iso/en/CombinedQueryResult.CombinedQueryResult?queryString=e-learning (accessed on 16 May 2007).
Kramer, B.J. 2000. Forming a federated virtual university through course broker middleware, in: *Proceedings: LearnTec 2000*. Heidelberg.
Lagrosen, S. & Seyyed-Hashemi, R. 2004. Examination of the dimensions of quality in higher education. *Quality Assurance in Education*, 12(2): 61–69.
Larson, M.R. & Bruning, R. 1996. Participant perceptions of a collaborative satellite-based mathematics course. *American Journal of Distance Education*, 10(1): 6–22.
McCollum, K. 1997. A professor divides his class in two to test value of online instruction. *Chronicle of Higher Education*, 43(24): A23.
Newton, R. 2003. Staff attitude to the development and delivery of e-learning. *New Library World*, 104(1193–2003): 412–425.
Osborne, M. & Oberski, I. 2004. University continuing education—the role of communications and information technology. *Journal of European Industrial Training*, 28(5): 414–428.
Owlia, M. S. & Aspinwall, E.M. 1996. A framework for the dimensions of quality in higher education. *Quality Assurance in Education*, 4(2): 12–20.
Quality Assurance Agency (QAA) (2002). QAA External Review Process for Higher Education in England: Operation Description (03/02). QAA, Gloucester.
Pajo, K. & Wallace, C. 2001. Barriers to the update of web-based technology by university teachers. *Journal of Distance Education*, 16(1): 70–84.
Pond, W.K. 2002. Distributed Education in the 21st century: Implication for quality assurance. *Online Journal of Distance Learning Administration*, 5(11). Available at www.westga.edu/~distance/ojdla/summer52/pond52.html (accessed on 11 May 2007).
Rovai, A.P. 2002. Building a sense of community at a distance, in: *International Review of Research in Open and Distance Learning*, April, ISSN: 1492–1831.
Sellani, R.J. & Harrington, W. 2002. Addressing administrator/faculty conflict in an academic online environment. *Internet and Higher Education*, 5: 131–145.
Tarr, M. 1998. Distance learning—bringing out the best. *Industrial and Commercial Training*, 30(3): 104–106.
Thomas, P. 1997. Teaching over the Internet, the future. *Computing and Control Engineering Journal*, 8(3): 136–142.
Training Press Releases (2006) "ISO/IEC standard benchmarks quality of e-learning" http://www.trainingpressreleases.com/newsstory.asp?NewsID=1767 (accessed on 16 May 2007).
Webb, J.P. 2000. Technology: a tool for the learning environment. *Campus-Wide Information Systems*, 18(2): 73–78.
Wilson, D.A. 2007. Tomorrow, tomorrow and tomorrow: the 'silent' pillar. *Journal of Management Development*, 26(1): 84–86.
Zhao, F. 2003. Enhancing the quality of online higher education through measurement. *Quality Assurance in Education*, 11(4): 214–221.

APPENDIX A: DIMENSIONS OF E-LEARNING QUALITY IN HIGHER EDUCATION

No	HE quality dimensions	E-Learning quality dimensions: ISO (2007)	US Institute benchmark (Gunasekaran et al., 2002, p. 48)	Webb (2000)	Newton (2003, p. 420)	Thomas (1997)	Zhao (2003)	Boticario and Guadioso (2000)	Alexander (2001)
1	Tangibles	Access	Password protection, encryption and other technical issues		Equivalent library and research opportunities		Adequacy of access in terms of technological infrastructure	Interactive	Technology development Technology network
2	Competence								Technical support
3	Attitude		Interaction and feedback	Learner's experience Timely remedial action for non-performers. Mentoring and guidance	Communication of requirements. Close personal interaction Counselling and advice	Support	Student satisfaction		Technical support to students
4	Content		Course development and design	Learner's experience			Effectiveness		
5	Delivery		Course delivery			Facilities for practical work (e.g., simulation). Enabling questions and discussions		Student participation. Individualized communication and quick access	

(*Continued*)

APPENDIX A: (*Continued*)

6	Reliability		Evaluation standards	Success criteria	Equally rigorous assessment as in campus-based learning; handling of plagiarism, authentication and online academic misconduct	Assessment			Market research
7	Globalization	Globalization e.g., language accessibility		Customize overall format for on-line delivery					
8	Creating communities of practice					Enabling questions and discussions		Community of practice—formal and informal e.g., "chat room"	
9	Developing e-learning vision, strategies and plans								Vision, strategies, policies (e.g., workload and time-release) and plans

Higher Education in the Twenty-First Century: Issues and Challenges – Al-Hawaj, Elali & Twizell (eds)
 ISBN 978-0-415-48000-0

Towards quality assurance and outstanding scientific research in higher education institutions in the Kingdom of Bahrain

Sadeq M. Al-Alawi
Higher Education Council, Kingdom of Bahrain

ABSTRACT: As knowledge is becoming the key driver for international competitiveness, the demand for quality education is increasing very quickly worldwide. During the rapid growth of higher education institutes in Bahrain since the year 2000, important sets of performance indicators are becoming extremely vital to ensure a high research quality rating as well as outstanding educational quality. To fulfil such requirements, new technologies should be implemented for a shift from teaching to learning. Furthermore, a clear vision, insight, and skill on the part of the administrators, faculty and students should be promoted. In addition, adequate assessment and evaluation should be created. In this presentation, several proposals towards a more effective system will be highlighted.

Higher Education in the Twenty-First Century: Issues and Challenges – Al-Hawaj, Elali & Twizell (eds)
© 2008 Taylor & Francis Group, London, ISBN 978-0-415-48000-0

Robust virtual reality environment for improving disability learning skills

Fatima Al Dhaen & H. Hussein Karam
Ahlia University, Kingdom of Bahrain

ABSTRACT: The expansion of 'care in the community' has highlighted the need for more effective educational and training media for people with learning disabilities. People who are physically disabled require a variety of access technology and learning depending on the nature of their disability. The intention of modern technology policy is to enable people with learning disabilities to have as much choice and control as possible over their lives, be involved in their communities, and make a valued contribution to the world at work. However, in order to achieve these aims, more effective educational and training media are needed as well as a Computer-Assisted Virtual Learning Environment (CVLE) for improving their skills and removing barriers that impede learning. Virtual reality (VR) possesses many qualities that give it rehabilitative potential for people with learning disabilities, both as an intervention and an assessment. By using computing technology for tasks such as reading and writing documents, communicating with others, and searching for information on the Internet, people with disabilities are capable of handling a wide range of activities independently. Still, they face a variety of barriers to computer use. These barriers can be grouped into three functional categories: barriers to providing computer input, interpreting output, and reading supporting documentation. This paper proposes a virtual reality (VR) framework as a solution for such challenges which helps to meet this need and enables access to information technologies by users with disabilities in a simple and reliable way. Evaluation and practical performance results will be performed. Discussion and an illustration of some experimental results give validation of the proposed technique.

Higher Education in the Twenty-First Century: Issues and Challenges – Al-Hawaj, Elali & Twizell (eds)

Prescriptions for educational quality assurance in the Kingdom of Bahrain

Martyn Forrest
Economic Development Board, Kingdom of Bahrain

ABSTRACT: Raising the quality of each educational level, be it elementary, secondary, vocational or higher education, underlies the educational reforms currently in progress in the Kingdom of Bahrain. An important complement to this multi-level approach is the creation of a new institution—the Polytechnic of Bahrain—to meet changing student and labour market demands. This paper sheds light on the crucial role of the Quality Assurance Agency (QAA) responsible for school evaluations, vocational educational inspections and university quality reviews. Some insights drawn from the experience of the Australian Universities Quality Agency serve as a backdrop for prescriptions on how to conduct quality assurance effectively in the Bahraini educational milieu.

Higher Education in the Twenty-First Century: Issues and Challenges – Al-Hawaj, Elali & Twizell (eds)
 ISBN 978-0-415-48000-0

Teaching and learning for the 21st century: A case for the Liberal Arts

Anastasia Kamanos
The Royal University for Women, Faculty of Education, Kingdom of Bahrain

ABSTRACT: We live in an era where everything is possible and nothing is certain. The changing nature of societal expectations has lead institutions of higher learning to respond to local, national and global pressures by re-shaping their academic programmes and practices. In the Middle East, the response can be observed in the proliferation of "American-style", liberal-arts-oriented independent undergraduate colleges and universities. However, a contradiction remains in that while forces external to higher education are dramatically reshaping it, institutional traditions and a misunderstanding of the "liberal arts" has meant that higher education has remained essentially the same. This paper examines the purpose of the "Liberal Arts" in the light of the "industry logic" and the "social institution logic" perspectives to explain how the Liberal Arts can best serve both views.

Higher Education in the Twenty-First Century: Issues and Challenges – Al-Hawaj, Elali & Twizell (eds)
 ISBN 978-0-415-48000-0

On the road to academic accreditation: The importance of research and balanced curricula

Rasha Abdulla
American University in Cairo, Egypt

ABSTRACT: Academic accreditation is no longer a luxury in today's academic world. It is a guarantee of an excellent quality education that is important for students, faculty members, and employers who utilize the services of an institution's graduates. For the academic institution, accreditation assures an excellent reputation, attracts potential students, and makes the institution more eligible for grants or sponsorships of various kinds. The road to accreditation or re-accreditation is challenging, exciting, and at times tiresome and confusing. But it is also a wonderful opportunity for schools to assess their strengths and weaknesses, rework their goals and objectives, and redirect their efforts to meet these goals and objectives. This presentation will tackle some of the important components of the educational process that are necessary for obtaining or maintaining accreditation. Specifically, the presentation will pay special attention to the importance of research as a component of a healthy educational institution, as well as the importance of having a balanced curriculum that maintains components of social sciences and general liberal arts education, in whatever major or specialization the student chooses. The paper outlines general steps on the road to acquiring accreditation from the main accrediting bodies, and how best to go about them. Then it will focus on some challenges that face institutions in the Arab world when applying for accreditation or re-accreditation. It will conclude with recommendations on how to make the accreditation process easier and more worthwhile and productive for all involved.

 ISBN 978-0-415-48000-0

Bibliographical notes of editors and authors

Abdulla Y. Al-Hawaj

President of Ahlia University

Dr Abdulla Y. Al-Hawaj, the President of Ahlia University, ranks in the top echelon of educational leaders in the Arabian Gulf. His incumbency in numerous managerial posts in prestigious organizations associated with educational development at a national level and his academic accolades from esteemed institutions, attest to Dr Al-Hawaj being in the vanguard of higher educational reform in the Kingdom of Bahrain. In recognition of Dr Al-Hawaj's trail-blazing contributions in the sphere of educational advancement in the Kingdom of Bahrain, he was appointed—by Royal decree in 2000 and 2006, respectively—not only as a member of the Supreme Committee for the Kingdom National Charter, in which he serves as the co-ordinator of its Educational Committee, but also as a member in the Higher Educational Council. Among his portfolio of managerial responsibilities, moreover, Dr Al-Hawaj, since 2001, has served as the President of the Bahrain Association for Academics. He holds a Ph.D. and an M.Sc. in Applied Mathematics from the University of Manchester (UK) and a B.Sc. (Hons) in Mathematics from the University of Kuwait. Dr Al-Hawaj launched his distinguished academic career at the University of Bahrain, in which he held the successive posts of Dean of Student Affairs, Chairman of the Department of Computer Science, and Chairman of the Department of Mathematics. He attained the rank of Professor of Mathematics at the University of Bahrain and continued to serve the University of Bahrain in that capacity until 2005. Dr Al-Hawaj's profound contributions to the fields of Mathematics, Computer Science, and Education through numerous research papers, books, media debates, and public discussions are vast. His unstinting dynamism, paradigm-shattering innovation and superlative achievement in the academic and social arena have accorded him a number of prestigious awards and honorariums. Dr Al-Hawaj has extensive experience in leading large multi-dimensional institutions. His signature use of pragmatic and innovative thinking in solving complex educational and business issues stems from his strong managerial and educational background. He is widely acclaimed as a pioneering architect of the private university in the Arabian Gulf and a leading proponent of excellence in Higher Education in the whole of the Middle East.

Wajeeh Elali

Professor Wajeeh Elali (M.A., MBA, Ph.D.) recently joined Ahlia University in Bahrain, after more than a decade teaching at McGill University, Canada. He has also taught at Northeastern University (USA), California State Polytechnic University (USA), and Concordia University (Canada). From 1981 to 1982 he was a Visiting Scholar at Harvard Business School. He has attended and presented numerous papers at international conferences and professional meetings in North America, Europe, the Middle East, Africa, and South Asia. His work has been published or accepted for publication in various refereed journals such as the Thunderbird International Business Review (USA), the International Journal of Business Governance and Ethics (UK), Financial Practice & Education (USA), the International Journal of Commerce & Management (USA), *Economia Internazionale* (Italy), Arab Studies Quarterly (USA), and the Journal of Accounting & Business Research (Canada). His current research focusses on corporate governance, business valuation, and international debt problems. Professor Elalai is the author of seven books in the field of corporate finance and productivity. Throughout his career, he has received a number of awards for excellence in teaching and research. In 2000, he received the Distinguished Teaching Award from McGill's Centre for Continuing Education. The award was given for outstanding graduate and undergraduate teaching. Among his prestigious recognitions, he received the Distinguished Teaching Award from McGill's Desautels Faculty of Management in 2003. He is also the recipient of the 1995 Distinguished Teaching Award from Concordia's John Molson School of Business.

Edward Henry Twizell

E.H. Twizell has *Professor Emeritus* status at Brunel University, U.K., where he was Professor of Mathematics until his retirement in September 2005. His research interests lie in the numerical solution of differential equations, with applications in real-world systems (mostly in the bio-medical sciences). A frequent visitor to Bahrain, he has co-supervised one Ph.D. student with Professor Abdulla Y. Al-Hawaj, and twenty-seven other Ph.D. students. Professor Twizell has published 170 research papers, and has written two books and edited three others. He is currently the Editor-in-Chief of the *International Journal of Computer Mathematics* (a Taylor & Francis publication).

The invited speakers

Chris Jenks

Vice-Chancellor and Principal of Brunel University

Professor Chris Jenks is Vice-Chancellor and Principal of Brunel University. He graduated from Surrey University in Sociology and Philosophy and completed his postgraduate work in Sociology at London University. Previously he was Professor of Sociology at the University of London. He is a prolific author and his distinguished publications include: Transgression—four volumes (2005); Childhood—three volumes (2005); Urban Culture—four volumes (2004); Subculture: Fragmentation of the Social (2004); Aspects of Urban Culture (2001); Transgression (2003); Culture: Critical Concepts—four volumes (2002); Culture (1993); Cultural Reproduction (1993); Visual Culture (1995); Childhood (1996); Core Sociological Dichotomies (1998); The Sociology of Childhood (1982); and Rationality, Education and the Social Organization of Knowledge (1976). His works have been translated into Chinese, Korean, Croatian, Polish, Danish, German, Turkish, Italian and Portuguese. He is an elected member of the Academy for Social Sciences and his academic interests span sociological theory, post-structuralism and heterology, childhood, cultural theory, visual and urban culture, and extremes of behaviour.

Don Betz

Chancellor of the University of Wisconsin—River Falls

Dr Don Betz is Chancellor of the University of Wisconsin—River Falls. He earned his Ph.D. from the University of Denver in International Studies and capped off his studies at Harvard University's Institute for Educational Management. Dr Betz is also a member of the Universities and Colleges Presidents' Climate Committee, the International Association of University Presidents, and is Vice-Chair of the International Education Committee. Previously, he was Provost and Vice-President for Academic Affairs and Professor of Political Science at the University of Central Oklahoma. Among his innovative activities in the sphere of international affairs, Dr Betz created and chaired the International Co-ordinating Committee on the Question of Palestine, a UN-affiliated non-governmental organization (NGO). A frequent writer and speaker, Dr Betz has addressed international, motivational and educational topics, and has worked with newspapers, radio and television. His life-long interest in global issues and his dedication to promoting cross-cultural understanding has led him to over 80 countries.

Amin A. Mahmoud

Former Minister of Culture, Hashemite Kingdom of Jordan

Dr Amin A. Mahmoud's distinguished career spans the halls of government and the towers of academia. Twice called upon to hold the cabinet post of Minister of Culture in the Hashemite Kingdom of Jordan, Dr Amin also undertook 15 years of academic leadership as President, successively, of the University of Petra and the University of Amman. Previously, he founded the Jordan University for Women in 1990 and served as its President for three years. In 1972, he earned his Ph.D. from Georgetown University in History and Government and subsequently served for

16 years as Professor of History and Political Science first at the University of Jordan and later at the University of Kuwait. Dr Mahmoud is the author of several books and more than 50 articles.

Marwan R. Kamal

President of Philadelphia University, Amman, Jordan

Dr Marwan R. Kamal is President of Philadelphia University in Amman, Jordan, and formerly held the post of President of the University of Bahrain. Subsequent to his completing his Ph.D. in Chemistry at the University of Pittsburgh (1961), Dr Kamal obtained an MBA (1968) from the same institution. He is also the current incumbent of the post of Secretary-General of the Association of Arab Universities in Jordan and is a Member of the Board of Trustees of Ahlia University (Bahrain) and Sharjah University (UAE). In the 1990s, among his other distinguished accomplishments, he was President of Yarmouk University (Irbid, Jordan) and Minister of Agriculture in the Hashemite Kingdom of Jordan. In the 1960s and 1970s, he served the King Fahd University of Petroleum and Minerals (Dhahran, Saudi Arabia) in a variety of academic and administrative positions and was a scientist at the Polymer Research Department of General Mills Corporation (Minnesota, USA). His many patents are registered in Europe and Japan and he is a frequent contributor to a wide variety of American scientific journals in diverse topical areas.

Syed Arabi Bin Syed Abdullah Idid

Rector of International Islamic University Malaysia (IIUM)

Dr Syed Arabi Idid, Professor of Communications, is the Rector of IIUM. Before his appointment, he served the same institution for five years as the Dean of the Research Centre (2001–2006). Previously, he held academic posts, including the deputy deanship of the Faculty of Arts & Sciences at the *Universiti Kebangsaan Malaysia* for 22 years. He earned his Ph.D. in communications from the University of Wisconsin (USA) and is a long-standing member of the World Public Opinion Research Association, the International Communication Association and the Asian Mass Communication and Information Centre. He is President of the Commonwealth Association for Education in Journalism and Communication. He has conducted several public opinion studies and is also involved in marketing research. Dr Idid's interests include conducting public opinion studies, public relations, and international communication.

Nabeel Al-Jama

General Manager Training & Career Development, Saudi Aramco

Mr Nabeel Al-Jama, General Manager, heads Saudi Aramco's Training & Career Development Organization in which he functions as the key executive in charge of planning, direction and guidance of a vast array of training programmes. Mr Al-Jama has played a pivotal role in priming current and prospective Aramco employees to handle Aramco's future business challenges and requirements. Holding a series of progressively responsible positions in central community services, before his promotion to GM in 2006, Mr Jama has more than two decades of corporate achievements in the service of Aramco. He joined Aramco in 1980. Mr Al-Jama holds a B.Sc. and an M.Sc. in community regional planning from King Fahd University of Petroleum and Minerals, Saudi Arabia.

Higher Education in the Twenty-First Century: Issues and Challenges – Al-Hawaj, Elali & Twizell (eds)
© 2008 Taylor & Francis Group, London, ISBN 978-0-415-48000-0

The delegates

Alparslan Açikgenç

Prior to his being appointed Vice-Rector, Fatih University, Istanbul, Turkey, in 2006, Dr Alparslan Açikgenç served the same distinguished university as Dean, Faculty of Arts & Sciences, for five years. Dr Açikgenç earned his Ph.D. in Philosophy from the University of Chicago in 1983. He is a prolific author in the spheres of ethics, logic, and philosophy of science and history of Islamic philosophy.

Faisal H. Al-Mulla

Dr Faisal H. Al-Mulla holds the rank of Associate Professor at the College of Education of the University of Bahrain and is Managing Editor of the Journal of Education & Psychological Sciences. He earned his Ph.D. in Education from the University of New Mexico, USA, in 1998 for which he was recognized with a Gold Medal from the Ministry of Education of the Kingdom of Bahrain.

Sofiane Sahraoui

Dr Sofiane Sahraoui, an Associate Professor in the School of Business & Management at the University of Sharjah, is currently a visiting researcher at Brunel University, UK. He received his Ph.D. in MIS from the University of Pittsburgh, USA, in 1994 and his research interests span e-Government, open source software, enterprise modelling and IT planning and change management.

Amer Al-Roubaie

Dr Amer Al-Roubaie is Dean of the College of Business & Finance at Ahlia University and holds a Ph.D. in Economics from McGill University in Montreal, Canada. He has had extensive teaching experience in North America, the Far East and the Middle East. In addition to having published well over forty articles on international economics, his principal research interest, Dr Al-Roubaie has written a book on globalization.

Kadom J.A. Shubber

Dr Kadom Shubber is a specialist in finance, investment, international business and general management with over twenty years teaching experience at several British universities including the University of Westminster. Focussing his research on North-South transfers of technology and international business, he earned his Ph.D. from Loughborough University of Technology, UK, in 1985. Dr Shubber is a Fellow of the Higher Educational Academy and of the Association of International Accountants. In addition, he is a long-standing member of the UK-based Chartered Institute of Management.

Purnendu Tripathi & Siran Mukerji

Dr Purnendu Tripathi, a faculty member in the Department of Business Administration, Arab Open University, KSA Branch—Riyadh, holds a Ph.D. in Management, an MBA in Marketing, and an M.Ed. in Distance Learning. He is co-author of a book on public administration, has written several research papers on management and distance learning.

Dr Siran Mukerji, a faculty member in the Department of Business Administration, Arab Open University, KSA Branch—Riyadh, holds a Ph.D. in Management and has the academic honour of having been designated a Jawahar Lal Nehru Scholar. She earned multiple Master's degrees spanning Distance Education, Public Administration and Human Resource Management, in which fields she has co-authored two books and written several research papers.

Deniz Saral

Dr Deniz Saral holds the post of Chairman, Business & Management Programmes, at Webster University, Geneva, Switzerland, and is Webster University—Europe's representative in the European Federation of Management Development (EFMD). Dr Saral obtained his Ph.D. in Business Administration from the University of Texas-Austin (USA) in 1975. He currently is active in international trade consulting and in executive management training.

Samia Costandi

Dr Samia Costandi is a teacher, researcher, writer and human rights activist who holds a Ph.D. in the Philosophy of Education from McGill University Faculty of Education in Montreal, Canada. While teaching and doing research at McGill during her 17 years of residency in Canada, she received the Helen Prize for Women in 1999 for merging her academic work with her community work and, in 2000, she obtained the Margaret Gillet Award from the McGill Centre for Teaching and Research on Women. Her areas of research and interest are, in addition to the philosophy of education: values in education, multi-cultural education, and mythology and education. Dr Costandi is of Palestinian origin and holds Canadian citizenship. In August 2005 he was invited to join the Royal University for Women in Manama, Bahrain, as *Programme Leader* in the Faculty of Education. She continues to teach and do research at RUW today.

Saravanan Nallaperumal & Shanthi Saravanan

Currently a staff member of the Computer Science and Engineering Department of Birla Institute of Technology, Bahrain, Mr Saravanan Nallaperumal is a doctoral candidate in computer science at M.S. University, Tirunveli, India. He also holds an M.Phil. in Mathematics and an M.Ed.Tech among his other credentials. His research interests span computational intelligence and e-learning and he is the author of several books and papers in these fields.

Currently a staff member of the Computer Science and Engineering Department of Birla Institute of Technology, Bahrain, Ms Shanti Saravanan is a doctoral candidate in the aforementioned institute in the field of artificial neural networks. She holds an M.Eng. from M.S. University, Madurai, India, in the field of computer science and has published several books and papers in the fields of neural networks and e-learning, her principal research interests.

Ahmed Y. Ali Mohamed

Dr Ahmed Y. Ali Mohamed is the Advisor to the University of Bahrain's President for Scientific Affairs. He earned a Ph.D. in Inorganic Chemistry from the University of Wales, UK, in 1982 and, among his multitude of affiliations, he is a Chartered Chemist, a member of the Royal Society of Chemistry and a member of the American Chemical Society. Apart from recent inquiries into inorganic aerosol particulate matter, his research interests are bifurcated between transitional metal complexes and solid state chemistry and he is the author of two books and over twenty-five technical articles in the field of chemistry.

Nabil Moussa

Professor Nabil Moussa received two B.Sc. degrees with honours in Electrical Engineering (1965) and Mathematics (1967) from Ain-Shams University, Egypt. He received the Ph.D. degree in Mathematics (1971) from Leipzig University, Germany. He taught many courses in Mathematics and Computer Science in several Arabic, American, and German Universities with 10 years as Head of the Mathematics Department and two years as Head of the Software Department at the Computer Centre of Essen University, Germany. Professor Moussa has been awarded more than 30 research and/or conference grants from international institutions. He is an Associate Editor and reviewer for International Journals such as AML, ELSEVIER Publishers, World Science and Engineering Academy and Society (WSEAS). Professor Moussa has published over 45 publications in refereed international journals and proceedings. He has also presented over 15 papers, plenary and invited lectures at International Conferences in more than 10 countries. He is the author of two text books.

Abdelkader Dagfous & Noor Al-Nahas

Dr Abdelkader Daghfous received his Ph.D. degree in Technology and Innovation Management from The Pennsylvania State University (USA). He received his B.S. and M.S degrees in Industrial Engineering from The Pennsylvania State University, and the University of Pittsburgh, respectively. He is currently an Associate Professor in the School of Business and Management at the American University of Sharjah, UAE. Prior to that, he taught at The Pennsylvania State University, Ahlia University, and the University of Bahrain, where he introduced several new courses in e-business, innovation, and knowledge management. His current research is in the areas of knowledge management, supply chain management, collaborative innovation, and the role of government-university-industry collaboration in regional economic development. He has published in renowned international refereed journals such as Research Policy, The Learning Organization, and Technovation. He has presented papers at various international conferences for academics and practitioners.

Noor Al-Nahas received her MBA and B.S. in Management Information Systems from the American University of Sharjah, UAE. She is currently working as a Technical Writer in the Information Systems Department at Global Information Technology in the UAE.

Abel Usoro & Abbas Abid

Dr Abel Usoro lectures in the School of Computing, University of the West of Scotland (formerly the University of Paisley, Scotland, UK). His current research interests are information systems, knowledge management and e-learning on which he has published widely. Dr Usoro is a full member of the prestigious Information Institute and British Computing Society and an international chair of several learned conferences—the ISOneWorld conference, the CITED (Conference on Information Technology and Economic Development) 2008, to name but two. He has also served as a Visiting Professor at Laurea University, Finland, and Anshan Normal University, PRC.

Dr Abbas Abid doubles as a lecturer in the School of Computing, University of the West of Scotland, and as a business training consultant. He obtained his Ph.D. in Business Administration in 1990 from the University of Glasgow and holds an M.Sc. (IT), which he subsequently earned from the University of Paisley in 2000. He has lectured both in British and Iraqi universities and his research interests centre on business information systems. He has published numerous scientific articles.

Jim Horn

Jim Horn joined Prince Mohammad Bin Fahd University (PMU) on 1 May 2006 as one of the Texas International Educational Consortium (TIEC) consultants to assist with the start-up phase at PMU. Mr Horn is not new to the Middle East since he most recently worked at Zayed University in the United Arab Emirates as the Director, Human Resources, during the period 2001–2005. Previously, Mr Horn, a University graduate from Simon Fraser University and a native Canadian, spent the past 30+ years of his career working in the Canadian university system. Mr Horn believes that universities play a very important role in any society and is committed to being part of the ongoing development of universities as an organization.

Anastasia Kamanos

Dr Anastasia Kamanos Gamelin is presently Programme Leader of the Liberal Arts and Associate Professor in the Education Faculty of the Royal University for Women in Bahrain. She has held a post-doctoral research fellowship at the University of Montreal's Centre for Ethnic Studies and has been a lecturer at McGill University's Faculty of Education. She obtained her Ph.D. (with Distinction) from McGill University, Canada. Her research and development activities have focussed on women in higher education, particularly on gender issues relating to women's education in the Middle East, and accordingly, she has been involved in establishing women's universities in Saudi Arabia and Bahrain.

Sadeq M. Al-Alawi

Dr Sadeq Al-Alawi is Director of Scientific Research at the Higher Educational Council of the Kingdom of Bahrain. He obtained a Ph.D. in Chemistry from Ohio University (USA) in 1997 and, for a decade, has held the rank of Assistant Professor at the University of Bahrain. Prior to commencing his academic career, Dr Al-Alawi worked as a senior chemical analyst at the Bahrain Petroleum Company (BAPCO). He earned numerous accolades from both the academic and industrial sectors and his research interests, in which he has published extensively, are diverse, spanning photo-chemistry of organic and inorganic compounds, surface catalysis and non-aqueous solutions.

Rasha Abdulla

Dr Rasha Abdulla is Assistant Professor of Journalism and Mass Communication at the American University in Cairo (AUC). She holds a Ph.D. in Communication from the University of Miami, USA. Dr Abdulla has taught at several American universities and has several prestigious academic and professional awards—most recently the Excellence in Research Award (2007) from AUC's School of Business, Economics and Communication. Her latest research delves into the impact of the Internet on the Arab world. Dr Abdulla is also a freelance writer/editor and has worked as a radio announcer at Radio Cairo's Overseas Department and the Local European Service.

Martyn Forrest

Dr Forrest is currently serving as the Executive Director, Education Reform, at the Economic Development Board (EDB) of the Kingdom of Bahrain. Dr Forrest was educated at the University of Western Australia and Oxford University and has held a number of senior positions in education in Australia including Vice-Chancellor of James Cook University, CEO of the Tasmanian Education Department and Deputy Chancellor of the University of Tasmania.

Fatima Al Dhaen & H. Hussein Karam

In 2002, Ms Fatima Al Dhaen, currently working as a graduate assistant at Ahlia University, obtained her B.Sc. in Multimedia Technology and Web Development from Napier University, UK. She is enrolled in Ahlia University's Master's degree programme in IT.

Dr Hussein Karam holds the rank of Associate Professor in Computer Science, College of Mathematics & IT, Ahlia University. He obtained his Ph.D. degree from Tokyo Institute of Technology, Japan, in 2000. Dr Karam has published several papers that reflect his diverse research interests, spanning computer graphics and animation, geometric modelling, multimedia, VR, CAD/CAM and fractal geometry.

Author index